HANDBOOK ON

1992

REFERENCE METHODS FOR

SOIL
ANALYSIS

Soil and Plant Analysis Council, Inc.

Handbook of Reference Methods for Soil Analysis

1. List of analytical methods for the analysis of soil for their chemical properties.

ISBN 0-9627606-1-7

Printed in the United States of America

TABLE OF CONTENTS

AAL- 3984

i

PREFACE

The first edition of the *Handbook on Reference Methods for Soil Testing* was published in 1974, and it was then revised in 1980. Both editions of the *Handbook* have proven to be a valuable reference for those engaged in research as well as those providing soil testing analytical services to farmers. The initial issue of 1000 copies of the *Handbook* in 1974 was exhausted by mid-1979. Therefore, the Council's first attempt to report on standard methods for soil testing had met with success, and the new revised addition was prepared for publication in 1980.

As prefaced in the 1974 publication, the Council intended to update and expand the *Handbook* to include additional methods and modifications of procedures developed since that date. That revision accomplished those intentions in several ways; namely,

1) the expression of results given on both a volume and weight basis.

2) addition of new procedures.

3) modifications of procedures found in the first *Handbook*.

The *Handbook* has changed over the years to reflect changing needs. A new method by Baker first appeared in the 1980 issue. Mehlich No. 2 and other methods were also added in 1980, and the Morgan procedure was dropped. In this issue, the Morgan procedure was placed back into the list of procedures, and the Mehlich No. 2 procedure has been replaced by Mehlich No. 3. Environmental concerns prompted the addition of sections on nitrate, heavy metals, and a method for estimation of organic matter that does not use dichromate. A section on quality assurance may be the most important addition. It addresses a most important factor effecting disagreement among laboratories after method selection and sample splitting. It also anticipates laboratory accreditation which seems to be in our future.

The major methods for managing plant nutrition currently in use in the United States and many other parts of the world are included. As with prior editions, no attempt has been made to

iv

include all methods for soil analysis, but to place those selected procedures into a standard format and to document the ramifications of each procedure. Continued research and changing needs will lead to changes in future editions. Comments and suggestion will be most helpful and are not only welcomed but are encouraged.

This third edition, like the ones before it, required the work of many people. All of the contributors listed in this edition reviewed, edited, or prepared the new sections. They added to the work of the contributors who assisted in the preparation of the earlier editions. Those individuals were: Fred Adams (Auburn University), Chris Chan (Ontario Ministry of Natural Resources, Canada), E. O. McLean (The Ohio State University), Adolf Mehlich (North Carolina Department of Agriculture), Doyle E. Peaslee (University of Kentucky), Kennard E. Pohlman (A&L Midwest Agricultural Laboratory), James R. Woodruff (Clemson University), and Zan R. Zwiep (A&L Great Lakes Agricultural Laboratory).

Nat B. Dellavalle
Chair 1990-1992

Council Officers and Executive Committee (1992)

Officers:
Nat B. Dellavalle - Chair
C. Owen Plank - Chair-Elect
J. Benton Jones, Jr. - Secretary-Treasurer

Executive Committee:

Donald J. Eckert	Byron Vaughan
Donald A. Horneck	Raymond Ward
Robert D. Munson	James R. Woodruff

Washington Liaison:
William C. White

CAST Representative:
Maurice Watson

CONTRIBUTORS

The following individuals assisted in the preparation of the *Handbook*, either by preparing specific methods or reviewing and editing methods that previously appeared in the 1980 version.

Dale E. Baker	Pennsylvania State University University Park, Pennsylvania
James R. Brown	University of Missouri Columbia, Missouri
Donald J. Eckert	The Ohio State University Columbus, Ohio
Clyde Evans	Auburn University Auburn, Alabama
Edward A. Hanlon	University of Florida Gainesville, Florida
Gordon V. Johnson	Oklahoma State University Stillwater, Oklahoma
J. Benton Jones, Jr.	Micro-Macro International, Inc. Athens, Georgia
Parviz N. Soltanpour	Colorado State University Fort Collins, Colorado
M. Ray Tucker	North Carolina Department of Agriculture-Agronomic Division Raleigh, North Carolina
Darryl D. Warnke	Michigan State University East Lansing, Michigan
Maurice E. Watson	REAL, OARDC Wooster, Ohio
Benjamin Wolf	Wolf's Agricultural Laboratory Ft. Lauderdale, Florida

ACKNOWLEDGEMENTS

Recognition is given for each the contributions to the Handbook, the list is given on the following page. In addition, special recognition is given to Mrs. Nickie Whitehead for editing and preparing most of the prepared copy, and to Mrs. Mani Ridgway for her scanning and preparation of the initial unedited text.

HANDBOOK ON REFERENCE METHODS FOR SOIL ANALYSIS

INTRODUCTION

For more than 30 years, soil testing has been widely used in this country as a basis for determining lime and fertilizer needs. To a large degree, the test procedures being used today by state and commercial soil testing laboratories originated in the late 1940s and early 1950s. In 1951 the Soil Test Work Group of the National Soil and Fertilizer Research Committee (1) surveyed the 50 state soil testing laboratories. The survey found that the laboratories employed 28 different extractants for determining phosphorus and 19 different extractants for determining potassium. Most of the methods and techniques were developed locally. Only the Morgan soil testing method (2) had wide use, particularly in the northeastern United States and in several western states.

In 1973, Jones (3) examined the soil test methodology in current use in the same laboratories and found that considerable standardization in methodology had taken place. Three methods—Bray P1, double acid, and Olsen—were used to determine extractable phosphorus. Similarly, three methods —neutral normal ammonium acetate, double acid, and Morgan—were used to determine extractable potassium. Yet, although the extractants used for determining soil phosphorus and soil potassium, respectively, are essentially three in number, laboratory techniques vary considerably. In 1980, a summary of soil testing methods found that the Morgan extraction procedure was less widely used. Two soil test work groups have published summaries of soil testing procedures in general use in the southeastern (4) and north central (5) regions of the United States. In addition, soil testing procedures have been published for Canada (6) and South Australia (7).

The present edition of the *Handbook* retains its initial purpose: to offer a standard laboratory technique manual for the more common soil testing procedures. In addition, the manual describes recent soil testing procedures, as proposed by Baker (8), Mehlich (9), and Soltanpour (10).

CHANGING ROLE AND NEEDS

The farmer or grower whose soil is properly analyzed and who follows the researcher's soil fertility recommendations carefully will ensure that all lime and fertilizer materials are correctly and economically used. Soil testing is big business for state, commercial, and fertilizer company laboratories. As larger numbers of cropped fields are being tested, more farmers and growers are coming to rely on soil test results. The increased interest in soil testing is due in part to the higher cost of fertilizer materials and to the desire of some to be environmentally correct in their use of agrichemicals. However, if soil testing is to be an effective means of evaluating the fertility status of soils, correct methodology is absolutely essential.

No single soil test is appropriate for all soils. Specific test procedures have been developed for certain soils, since one research method cannot be used indiscriminately over a wide range of soil-crop conditions. However, there is some interest in developing more universal test procedures that can be adapted to a wide range of soil-crop conditions. At present, a number of procedures are used to determine soil pH, lime requirement, and level of extractable nutrient elements. Increasingly, researchers see a more significant role for soil testing as a monitoring tool and as a means of estimating lime and fertilizer needs on a regional or national scale.

The variance in methodology introduces differences in test results which lead to confusion among farmers and farmer advisors. Failing to realize that there is no single soil test for all soils, these workers may unknowingly send soil samples to a laboratory which does not employ tests applicable to their particular soil type. Also, some laboratories fall to realize that their particular soil testing procedures may not be applicable to all the soils they receive. The result may well be a faulty interpretation and a recommendation which produces an imbalance problem for the farmer.

Although the number of extractants has greatly narrowed in the last 20 years, techniques vary markedly. In one study, Jones (3) noted that among those who used the Bray P1 extraction process for determining phosphorus, soil-to-solution ratios varied from 1:6.7 to 1:10 and shaking times from 40 seconds to 5 minutes. Bray's original procedure called for a 1:7 soil-to-solution weight ratio with a 40-second shaking time.

Similar differences exist for other test methods. In the extraction of potassium by neutral normal ammonium acetate, soil-to-solution ratios have varied from 1:3 to 1:10 with various shaking times. Soil pH is determined in either 1:1 or 1:2 soil-water mixes, in saturated pastes, or in dilute salt solutions (3). Some laboratories weigh samples; others scoop to a known volume or an estimated weight. Thus the amount of soil extracted can alter the test result and affect the interpretation. For those who rely on more than one laboratory to provide soil tests for farmers in various regions of the United States, these differences in methodology create an almost intolerable situation.

The various ways of expressing results cause confusion, too, particularly for those who work across state lines or who submit samples to more than one laboratory. The volume measurement, pounds per acre, and the weight measurement, parts per million, are commonly used expressions denoting the level for an extracted element. As all too few realize, an error is introduced into the soil test result when the volume measurement is used for weighed laboratory samples, or when volume- measured sample final results are expressed as parts per million. To the unsuspecting, differing results for the same soil may be caused by an erroneous mathematical calculation. Mehlich (11,12) addresses these problems in two practical discussions of volume and weight concepts.

Soil test results are also commonly expressed in terms of a sufficiency range, the usual designation denoting a test result as either "low," "medium," or "high." Such a coded system is discussed by Cope (13). A detailed discussion on soil test interpretation may be found in ASA Special Publication Number 29 (14) and SSA publication, *Soil Testing and Plant Analysis* (revised edition) (15).

The actual method of soil testing is relatively unimportant, compared to the interpretation of the soil test result, particularly the lime and fertilizer recommendations. If the laboratory test result is well correlated with crop response or yield, the method of obtaining the test result is of minor concern. However, in too many instances laboratory modifications of an already calibrated soil test are made without recalibration. For example, many laboratories use Bray's P1 phosphorus test calibration and inter-pretation data but have modified Bray's original laboratory procedure. In these instances, the modification alters the test result sufficiently to invalidate the interpretation as originally proposed. The result is a faulty interpretation and an improper phosphorus fertilizer recommendation.

REFERENCE METHODS

The need to standardize soil testing procedures and methods is more apparent today, although there is little unanimity on the subject. Several groups are evaluating various testing methods and setting parameters for each laboratory procedure. Much work is being done by regional research committees on soil testing.

A number of scientific and industrial societies have been engaged in developing and publishing reference methods of analysis. The Association of Official Analytical Chemists (AOAC), which was organized in 1884, is the oldest of these societies in the United States. The American Society of Testing Materials (ASTM), the American Public Health Association, and, more recently, the Intersociety Committee have been engaged in researching and publishing reference methods of analysis for a wide variety of substances, including biological materials. The Council on Soil Testing and Plant Analysis was organized in 1970 to research and prepare reference methods for soil testing.

The potential and changing role of soil testing demands reference test methods. The growing interest in the environment, the concern by some about overdosing our soils with fertilizer, and the need for care in using fertilizer materials in short supply demand more uniformly applied test methods. Standardization of methodology is indeed necessary if soil testing is to be used as a valid monitoring tool. In the near future, regulating agencies may dictate the methods of soil analyses, much as various governmental agencies have required AOAC methods for analyses of fertilizers, lime, and other substances. If soil testing is to become a scientific endeavor, standardization of processes is essential.

Much of the research being published today on crop production contains a considerable amount of soil test data which is of doubtful value. Frequently, the publication does not include references to a particular procedure or provide a detailed description of the method under examination. An article may merely refer to a particular soil test by name without listing the steps of the testing procedure. To address this discrepancy, the *Handbook* provides a complete description of many different soil test procedures, thereby increasing the value of the published information.

SOIL SAMPLE EXCHANGES

A number of groups, including the Council, have engaged in round-robin soil sample exchanges. Participating laboratories receive specially prepared soil samples and solution standards for analysis. Each laboratory uses its own methodology to compare its results to those of other laboratories. Analyses are performed for water pH; for extractable phosphorus by the Bray P1, double acid, and Morgan extractants; and for extractable potassium, calcium, and magnesium by the neutral normal ammonium, acetate, double acid, and Morgan extractants. Tests for organic matter content and extractable zinc have also been the focus of several round-robin studies. The Council exchanges have involved some 45 soil testing laboratories, all located in the eastern half of the United States. In general, most laboratories, irrespective of technique, agree closely on water pH determined on these exchange soil samples. However, large variations (10% to 30%) have been obtained for most of the extractable element determinations.

These results have been carefully evaluated to determine the nature and source of the variance. Most exchanges emphasize the need for a higher level of analytical skill in the laboratory and for a standardization of techniques among laboratories using the same soil test procedure. Although poor analytical techniques account for some of the variance, it is equally apparent that differences in sample size, soil-to-solution ratio, extraction vessel size and shape, shaking speed, and length of shaking period are contributing factors (16). The soil sample exchanges and studies clearly point to the need for uniformity.

A survey indicated that many managers of state soil testing laboratories, as well as some managers of commercial soil testing laboratories, do not want to standardize soil testing procedures. Some technicians see standardization as an unnecessary cause of change and regulation, believing that uniformity will lead to the permanent establishment of a test methodology that offers little opportunity for innovation. Similarly, some feel that standardization may discourage research into the development of new and improved testing methods. Others see the importance of standardizing laboratory procedures but want to leave the interpretation and recommendation phase intact.

These views relating to the standardization of soil test methods need careful study. Much of the stir today in soil testing seems to have its roots among those who want to see "science" added to the "art" of soil testing. For those who see this need, the development of reference methods is a first and most important step.

GENERAL CONSIDERATIONS

The *Handbook of Reference Methods for Soil Testing* defines the parameters of several of the more frequently used soil tests. Considerable effort has been dedicated to presenting sufficient details of each method so that the reader would not have to refer to other sources. Pertinent references are provided, as appropriate.

In most instances, a dual system of weighed and/or volume measured samples is presented. This rationale is necessary in cases in which the original method carefully specified a weight of sample or volume of known or assumed specific weight. The reader may refer to Mehlich (11, 12) for additional information on volume-weight considerations; to Peck (17) for more details on scoop design and use.

Sample Scooping

In this *Handbook,* scoop size is based on an assumed "average" volume-weight of prepared sample air-dry, 10-mesh, 2-mm soil. The typical soil prepared for analysis, as described in this manual, has an assumed weight-to-volume ratio of 1.18 for silt loam and clay textured soils, and 1.25 for sandy soils. Therefore, those soil test procedures adapted to a particular textured type soil will designate scoop volumes that match the assumed weight-to-volume ratio:

Silt loam and clay textured soils		Sandy soils	
weight	scoop size	weight	scoop size
g	cm^3	g	cm^3
2.5	1.70	5.0	4.0
5.0	4.25		
10.0	8.50		

Scoops are of a fixed volume and do not necessarily yield an estimated or assumed weight. However, when the volume weight of a soil sample is known, a specific volume of that soil can be scooped to give an estimated weight.

Since an accurate measure for scooped samples is essential, scoop design and the scooping technique become important. Peck (17) has devoted considerable study to the scooping process. He found that for best results, the diameter of the scoop should be approximately twice the depth.

The technique for scooping is as follows:

1. Dip the scoop into the center of the soil sample deep enough to fill the scoop heaping full.

2. Strike the handle near the scoop itself three times with a spatula.

3. Level the scoop with a sharp edge.

Soil samples collected by following this procedure will be within 2% to 3% of the same volume or estimated weight. Scooping of soil samples has been found to give results comparable to weighed samples in repeated analyses of the same soil sample.

SUMMARY

This revised edition of the 1974 *Handbook* is dedicated to the improvement of the soil testing technique as it is applied in the laboratory. Each procedure given has been carefully researched, and the procedures are presented so that they can be easily followed. Using the described procedures under actual laboratory conditions will, it is hoped, lead to additional improvements in technique.

REFERENCES

1. Nelson, W. L., J. W. Fitts, L. D. Kardos, W. T. McGeorge, R. Q. Parks, and J. Fielding Reed. 1953. Soil testing in the United States, 0-979953. National Soil and Fertilizer Research Committee, US Government Printing Office, Washington, DC.

2. Lunt, H. A., C. L. W. Swanson, and H. G. M. Jacobson. 1950. The Morgan soil testing system. Conn. Agr. Exp. Sta. Bull. 541:1-60.

3. Jones, Jr., J. B. 1973. Soil testing in the United States. Comm. Soil Sci. Plant Anal. 4:307-322.

4. Sabbe, W. E., and H. L. Breland. 1974. Procedures used by state soil testing laboratories in the southern region of the United States. Southern Cooperative Series Bull. No. 190. 23 p.

5. Dahnke, W. C. (ed.). 1975. Recommended chemical soil test procedures for the north central region. North Central Regional Pub. 221. North Dakota Agr. Exp. Sta. Bull 499. 23 p.

6. McKeague, J. A. (ed.). 1978. Manual of soil sampling and methods of analysis (2nd ed.). Canadian Society of Soil Science, Ottawa, Ontario, Canada. 211 p.

7. Heanes, D. (ed.). 1977. Laboratory methods of soil and plant analysis (2nd ed.). Soil Conservation Branch Report S5/77. Department of Agriculture and Fisheries, South Australia. 150 p.

8. Baker, Dale E. 1973. A new approach to soil testing: II. Ionic equilibrium involving H, K, Ca, Mg, Mn, Fe, Cu, Zn, Na, P, and S. Soil Sci. Soc. Am. Proc. 37:537-541.

9. Mehlich, A. 1978. New extractant for the soil test evaluation of phosphorus, potassium, sodium, calcium, magnesium, and manganese. Comm. Soil Sci. Plant Anal. 9(6):477-492.

10. Soltanpour, P. N., and A. P. Schwab. 1977. A new soil test for simultaneous extraction of macro- and micro-nutrients in alkaline soils. Comm. Soil Sci. Plant Anal. 8(3): 195-207.

11. Mehlich, A. 1972. Uniformity of expressing soil test results: A case of calculating results on a volume basis. Comm. Soil Sci. Plant Anal. 3:417- 424.

12. Mehlich, A. 1973. Uniformity of soil test results as influenced by volume weight. Comm. Soil Sci. Plant Anal. 4:475-486.

13. Cope, Jr., J. T. 1972. Fertilizer recommendations and computer programs key used by the soil testing laboratory. Auburn Exp. Sta. Circular 176. 55p.

14. Peck, T. R., J. T. Cope, Jr., and D. A. Whitney (ed.). 1977. Soil testing: Correlating and interpreting the analytical results. ASA Special Pub. No. 19. Madison, WI. 117 p.

15. Walsh, L. M., and J. D. Beaton (eds.). 1973. Soil testing and plant analysis (rev. ed.). Soil Science Society of America, Madison, WI. 491p.

16. Grava, John. 1975. Causes for variation in phosphorus soil tests. Comm. Soil Sci. Plant Anal. 6(2): 129-138.

17. Peck, T. R. 1975. Standard soil scoop, pp. 4-5. IN: W. C. Dahnake (ed.), Recommended chemical soil test procedures for the north central region. North Central Regional Pub. No. 221. North Dakota Agr. Exp. Sta. Bull. 449.

SUMMARY TABLES

The following tables provide the user of this handbook with an overview of the basic parameters for several soil test methods described in more detail in the later sections. The various parameters given also provide a means of quickly comparmg one method to another. However, the user should refer to the more detailed descriptions for each of the soil test methods when conducting a particular soil test in the laboratory.

Table 1. Soil pH in Water, 0.01M Calcium Chloride, and Buffer pH by SMP and Adams-Evans Buffers, and Soil Paste pH

Soil Test	Sample Size	Extraction Reagent	Volume Used	Shaking Time	Method of Determination	References
Soil Water pH (for SMP buffer)	5.0 g (4.25-cm³)	Water	5 mL	10 min. with intermittent stirring	Read pH to nearest 0.1 with glass electrode pH meter calibrated with pH 4.0 and 7.0 buffers	Method of Soil Anal., No. 9:914-926. Am. Soc. Agron. (1965); Soil Test & Plant Anal., pp. 77-95. Soil Sci. Soc. Am. (1973)
Soil Water pH (for Adams-Evans buffer)	10-cm³	Water	10 mL	10 min. with intermittent stirring	Read pH to nearest 0.1 with glass electrode pH meter calibrated with pH 4.0 and 7.0 buffers	
Soil pH in 0.01M CaCl₂	5.0 g (4.25-cm³)	0.01M CaCl₂	5 mL	30 min. with intermittent stirring	Read pH to nearest 0.1 with glass electrode pH meter calibrated with pH 4.0 and 7.0 buffers	Soil Sci. Soc. Am. Proc. 19:154-167 (1955)
SMP Buffer pH	5.0 g (4.25-cm³) soil in 5 mL water	SMP Buffer	10 mL	Shake 10 min., intermittent stirring for 20 min.	Read pH to nearest 0.1 with glass electrode pH meter calibrated with pH 4.0 and 7.0 buffers	Soil Sci. Soc. Am. Proc. 24:274-277 (1961)
Adams-Evans Buffer pH	10-cm³ soil in 10 mL water	Adams-Evans Buffer	10 mL	Shake 10 min., intermittent stirring for 20 min.	Read pH to nearest 0.5 with glass electrode pH meter calibrated to read pH 8.0 in mixture of 10 cc buffer reagent and 10 cc water.	Soil Sci. Soc. Am. Proc. 26:355-357 (1962)
Mehlich Buffer pH	10-cm³ soil in 10 mL water	Mehlich Buffer	10 mL	60 min. with intermittent stirring.	Read pH to nearest 0.05 with glass electrode pH meter calibrated to read pH 6.6 in mixture of 10 cc buffer reagent and 10 cc water.	Comm. Soil Sci. Plant Anal. 7(7):637-652 (1976)
Soil Paste pH	About 150 g	Water	Varies with texture	1 hour	Read pH to nearest 0.1 with glass electrode pH meter calibrated with pH 4.0 and 7.0 buffers.	Diagnosis and improvement of saline and alkali soils. USDA

Table 2. Soil Test Phosphorus Methods

Parameter	Method				
	Mehlich No. 1 (Double Acid)	Bray P1	Mehlich No. 2	Olsen	AB-DPTA
Adaptability Limits	Sandy soils, acid, low in CEC	Acid soils with moderate CEC	All soils	Alkaline soils	Alkaline soils
Sample Size, Weight (volume)	5 g (4-cm³)	2 g (1.70-cm³)	2.5-cm³	2.5 g (2-cm³)	10 g (8.5-cm³)
Volume of Extractant, mL	25	20	25	50	20
Extracting Solution	0.05N HCl in 0.025N H₂SO₄	0.03N NH₄F in 0.025N HCl	0.2N HOAc in 0.015 N NH₄F in 0.2N NH₄Cl in 0.012N HCl	0.5N NaHCO₃ at pH 8.5	1M NH₄HCO₃ 0.005M DTPA at pH 7.6
Shaking Time, minutes	5	5	5	30	15
Shaking Action and Speed	Reciprocating 180+ oscillations/min.	Reciprocating 180+ oscillations/min.	Reciprocating 180+ oscillations/min.	Reciprocating 180+ oscillations/min.	Reciprocating 180+ oscillations/min.
Method of P Determination in Extract	Molybdenum Blue	Molybdenum Blue	Molybdenum Blue	Molybdenum Blue	Molybdenum Blue
Range in Soil P Conc. without Dilution, kg/ha	2-100	2-250	2-200	2-200	2-100
Sensitivity, kg/ha	1	1	1	1	2
Primary Reference	North Carolina Soil Test Div. mimeo (1953)	Soil Sci. 59:39 (1945)	Comm. Soil Sci. Plant Anal. 9 (6):477 (1978)	USDA Circular 939 (1954)	Comm. Soil Sci Plant Anal. 8:195 (1977)

Table 3. Soil Test Potassium Methods

Parameter	Mehlich No. 1 (Double Acid)	1N NH$_4$OAc pH 7.0	Mehlich No. 2	Water	AB-DPTA
			Method		
Adaptability Limits	Sandy soils, acid, low in CEC	Wide range of soils	Wide range of soils	Wide range of soils	Alkaline soils
Sample Size, Weight (volume)	5 g (4-cm^3)	5 g (4.25-cm^3)	2.5-cm^3	5 g (4.25-cm^3)	10 g (8.5-cm^3)
Volume of Extractant, mL	25	25	25	25	20
Extracting Solution	0.05N HCl in 0.025N H$_2$SO$_4$	1N NH$_4$OAc at pH 7.0	0.2N HOAc in 0.015 N NH$_4$F in 0.2N NH$_4$Cl in 0.012N HCl	Pure water	1M NH$_4$HCO$_3$ 0.005M DTPA at pH 7.6
Shaking Time, minutes	5	5	5	30	15
Shaking Action and Speed	Reciprocating 180+ oscillations/min.	Reciprocating 180+ oscillations/min.	Reciprocating 180+ oscillations/min.	Reciprocating 180+ oscillations/min.	Reciprocating 180+ oscillations/min.
Method of K Determination in Extract	Flame Emission Spectroscopy	Flame Emission Spectroscopy	Flame Emission Spectroscopy	Flame Emission Spectroscopy	Atomic Adsorption
Range in Soil K Conc. without Dilution, kg/ha	50-400	50-1000	50-1000	50-500	5-750
Sensitivity, kg/ha	5	5	5	5	1
Primary Reference	North Carolina Soil Test Div. mimeo (1953)	Soil Sci. 59:13 (1945)	Comm. Soil Sci. Plant Anal. 9 (6):477 (1978)	Am. Soc. Agron. Pub. No. 9:935-945 (1965)	Comm. Soil Sci Plant Anal. 8:197 (1977)

Table 4. Soil Test Calcium Methods

Parameter	Method			
	Mehlich No. 1 (Double Acid)	1N NH$_4$OAc at pH 7.0	Mehlich No. 2	Water
Adaptability Limits	Sandy soils, acid, low in CEC	Wide range of soils	Wide range of soils	Wide range of soils
Sample Size	5 g (4-cm^3)	5 g (4.25-cm^3)	2.5-cm^3	5 g (4.25-cm^3)
Volume of Extractant, mL	25	25	25	25
Extracting Solution	0.05N HCl in 0.025N H$_2$SO$_4$	1N NH$_4$OAc at pH 7.0	0.2N HOAc in 0.015 N NH$_4$F in 0.2N NH$_4$Cl in 0.012N HCl	Pure water
Shaking Time, minutes	5	5	5	30
Shaking Action and Speed	Reciprocating 180+ oscillations/min.	Reciprocating 180+ oscillations/min.	Reciprocating 180+ oscillations/min.	Reciprocating 180+ oscillations/min.
Method of Ca Determination in Extract	Atomic Absorption Spectroscopy	Atomic Absorption Spectroscopy	Atomic Absorption Spectroscopy	Atomic Absorption Spectroscopy
Range in Soil Ca Conc. without Dilution, kg/ha	120-1200	500-2000	500-2000	150-1000
Sensitivity, kg/ha	10	10	10	10
Primary Reference	North Carolina Soil Test Div. mimeo (1953)	Soil Sci. 59:13 (1945)	Comm. Soil Sci. Plant Anal. 9(6):477 (1978)	Am. Soc. Agron. Pub. 9:935-945 (1965)

Table 5. Soil Test Magnesium Methods

Parameter	Method			
	Mehlich No. 1 (Double Acid)	1N NH$_4$OAc at pH 7.0	Mehlich No. 2	Water
Adaptability Limits	Sandy soils, acid, low in CEC	Wide range of soils	Wide range of soils	Wide range of soils
Sample Size	5 g (4-cm^3)	5 g (4.25-cm^3)	2.5-cm^3	5 g (4.25-cm^3)
Volume of Extractant, mL	25	25	25	25
Extracting Solution	0.05N HCl in 0.025N H$_2$SO$_4$	1N NH$_4$OAc at pH 7.0	0.2N HOAc in 0.015 N NH$_4$F in 0.2N NH$_4$Cl in 0.012N HCl	Pure water
Shaking Time, minutes	5	5	5	30
Shaking Action and Speed	Reciprocating 180+ oscillations/min.	Reciprocating 180+ oscillations/min.	Reciprocating 180+ oscillations/min.	Reciprocating 180+ oscillations/min.
Method of Mg Determination in Extract	Atomic Absorption Spectroscopy	Atomic Absorption Spectroscopy	Atomic Absorption Spectroscopy	Atomic Absorption Spectroscopy
Range in Soil Mg Conc. without Dilution, kg/ha	40-360	50-500	50-500	50-500
Sensitivity, kg/ha	5	5	5	5
Primary Reference	North Carolina Soil Test Div. mimeo (1953)	Soil Sci. 59:13 (1945)	Comm. Soil Sci. Plant Anal. 9(6):477 (1978)	Am. Soc. Agron. Pub. 9:935-945 (1965)

Table 6. Soil Test Sodium Methods

Parameter	Method			
	Mehlich No. 1 (Double Acid)	1N NH$_4$OAc at pH 7.0	Mehlich No. 2	Water
Adaptability Limits	Sandy soils, acid, low in CEC	Wide range of soils	Wide range of soils	Wide range of soils
Sample Size	5 g (4-cm^3)	5 g (4.25-cm^3)	2.5-cm^3	5 g (4.25-cm^3)
Volume of Extractant, mL	25	25	25	25
Extracting Solution	0.05N HCl in 0.025N H$_2$SO$_4$	1N NH$_4$OAc at pH 7.0	0.2N HOAc in 0.015 N NH$_4$F in 0.2N NH$_4$Cl in 0.012N HCl	Pure water
Shaking Time, minutes	5	5	5	30
Shaking Action and Speed	Reciprocating 180+ oscillations/min.	Reciprocating 180+ oscillations/min.	Reciprocating 180+ oscillations/min.	Reciprocating 180+ oscillations/min.
Method of Na Determination in Extract	Flame Emission Spectroscopy	Flame Emission Spectroscopy	Flame Emission Spectroscopy	Flame Emission Spectroscopy
Range in Soil Na Conc. without Dilution, kg/ha	10-200	10-250	10-200	10-200
Sensitivity, kg/ha	2	2	2	2
Primary Reference	North Carolina Soil Test Div. mimeo (1953)	Soil Sci. 59:13 (1945)	Comm. Soil Sci. Plant Anal. 9(6):477 (1978)	Am. Soc. Agron. Pub. 9:935-945 (1965)

Table 7. Soil Test Zinc Methods

Parameter	Method			
	Mehlich No. 1 (Double Acid)	0.1N HCl	DPTA	AB-DTPA
Adaptability Limits	Sandy soils, acid, low in CEC	Acid soils	Wide range of soils	Alkaline soils
Sample Size	5 g (4-cm^3)	5 g (4.25-cm^3)	10 g (8.5-cm^3)	10 g (8.5-cm^3)
Volume of Extractant, mL	20	20	20	20
Extracting Solution	0.05N HCl in 0.025N H$_2$SO$_4$	0.1N HCl	0.005M DTPA, 0.1M TEA, 0.01M CaCl$_2$	1M NH$_4$HCO$_3$, 0.005M DTPA at pH 7.6
Shaking Time, minutes	5	30	120	15
Shaking Action and Speed	Reciprocating 180+ oscillations/min.	Reciprocating 180+ oscillations/min.	Reciprocating 180+ oscillations/min.	Reciprocating 180+ oscillations/min.
Method of Zn Determination in Extract	Atomic Absorption Spectroscopy	Atomic Absorption Spectroscopy	Atomic Absorption Spectroscopy	Atomic Absorption Spectroscopy
Range in Soil Zn Conc. without Dilution, kg/ha	2-30	2-20	0.5-20	0.5-35
Sensitivity, kg/ha	1	1	0.5	0.5
Primary Reference	Comm. Soil Sci. Plant Anal. 1:35 (1970)	Soil Sci. Soc. Am. Proc. 32:543 (1968)	Soil Sci. Soc. Am. Proc. 33:62 (1969)	Comm. Soil Sci. Plant Anal. 8:195 (1977)

QUALITY ASSURANCE PLANS FOR AGRICULTURAL TESTING LABORATORIES

INTRODUCTION

It is the intent of every agricultural testing laboratory, regardless of size, to provide reliable testing results to its clientele. Each measurement made by the laboratory is associated with some degree of uncertainty. This uncertainty is introduced into every process step of an analysis, because the analysis itself is designed to estimate unknown values. To meet the goal of "reliable testing results," a laboratory must have some way of proving its credibility. This section will discuss one method of providing credibility: a quality assurance (QA) program.

Here the term "process" refers to a discrete part of a measurement methodology. "Analysis" refers to the result of the overall methodology. Often, agricultural measurements consist of several process steps within an analysis. For example, for soil sampling, the processes of sample preparation, sample extraction, analytical measurement, data reduction, and final report might describe an analysis. Each process step introduces some uncertainty into the final data produced from the analysis.

QUALITY ASSURANCE IN THE LABORATORY

Providing credibility through a QA program is something that at first may seem to be counterproductive. That is, the more testing a laboratory does to prove that unknown measurements are correct, the fewer unknown samples it can complete in a given time period.

In fact, once the QA process is fully implemented throughout the laboratory, the laboratory may actually be completing more unknowns. The reason for this apparent contradiction is twofold. First, the accuracy and precision will be known for the system and, therefore, can be demonstrated to the client with fewer QA samples. Second, when the system of testing is within statistical control (*i.e.*, data from the QA program meet preselected statistics), fewer unknowns will have to be rerun. Thus, the efficiency of the process is increased and the total cost for completing unknown samples decreases. (See the section on statistical control.)

The ability to reproduce a value time and again (precision) or the ability to attain a known result (accuracy) indicates that the testing process operates within known uncertainty levels. If accuracy and precision levels can be demonstrated to be within the goals of both the client and the internal laboratory guidelines over a sufficiently long time period and testing range, the process is much more likely to continue to produce quality data in the future. Only a few tests need be completed to demonstrate continued accuracy and precision of the process.

Uncertainty can be introduced into the measurement process from many sources. Technical staff work within the uncertainty of the process. When the uncertainty of the process is known, then digressions from the normal level of uncertainty can be detected easily and appropriate changes made before additional testing, of inferior quality, continues. Technicians can detect inferior results immediately, spending valuable time more wisely (recalibrating, etc.) rather than being told about the error later by the manager and having to repeat previous testing. The reduced workload also results in reduced waste which, in turn, keeps operating costs low.

Knowing How a Laboratory Operates

Preparing sample analyses is the real purpose of a laboratory. Knowing how the process really works, a leader (manager) can identify possible problem areas within a process. It should be the goal of an effective leader to remove any stumbling blocks to quality which the technical staff face. The technical staff work within the process, while the leader works on the process itself (Deming, 1986).

Figure 1 describes the "chain of command" in a typical laboratory facility. However, the diagram does not explain how the work is actually accomplished, nor does it detail who does the work. Instead, it implies that the director initiates all processes. The implication that the technical staff contribute little to the laboratory other than labor is inherent in such a diagram. Yet it is the training and expertise of the technical staff that contribute directly to the success of the laboratory. For this reason, these diagrams are of little benefit in improving the internal processes of the laboratory, nor do they contribute directly to the QA process of the laboratory.

The internal processes or work paths can better be described for QA improvement if diagrams such as Figure 2 are

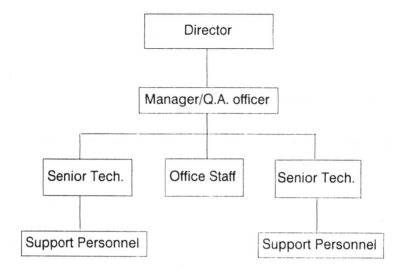

Figure 1. Chain of Command for an Analaytical Laboratory.

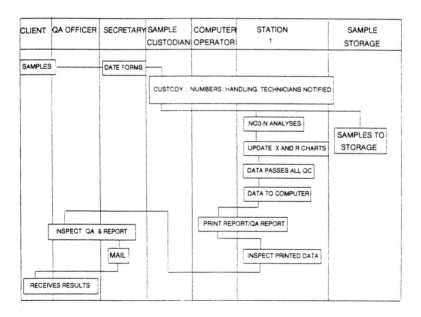

Figure 2. Internal Process or Work Paths in an Analytical Laboratory.

developed. The information presented in Figure 2 is a partial diagram of the sample process, showing only the pathway for nitrates (Station I). The main contributors to the process are listed across the page in columns, while the work completed by the contributors is listed in the various shaped borders. Each horizontal line indicates that one contributor transfers his/her work to another contributor. In other words, one is acting as a supplier and the other as a customer.

It is here that the QA management perspective can be promoted. If the customer has to do additional work (inspections, data entry, etc.) to bring the "product" of the supplier to standards, a process change at the supplier side may be needed. Additionally, if the supplier is unaware of the needs of the customer, further QA problems can be introduced. Quality begins with the supplier. It is the leader's responsibility to address the process, on both supplier and customer sides, to remove these impediments to quality.

It is often useful to have the technical staff create the initial version of these diagrams and to discuss the results in staff meetings. Since the technical staff works within each of these processes, who is better able to explain the process as it actually exists? When similar diagrams were first drafted at the University of Florida, Institute of Food and Agricultural Sciences, Analytical Research Laboratory, it was found that a rerouting of the work path eliminated some unneeded steps and reduced the overall workload by 5% for one technician. This change, initiated after the new process had been tested, had no measurable effect on the quality of data originating from the process.

Once the process has been diagrammed to the satisfaction of all contributors, the standard operating procedures (SOP) within a process can be identified and prepared. The diagrams must take into account all sample handling, data transfer, and paperwork within a process to insure that the manager can address all possible points within the process needing QA activity. The written SOP text contains only those parts of the process that need further explanation not provided in the diagram (see the section on documentation, below).

STATISTICAL CONTROL

The uncertainty of a process (and its major contributing components) must be quantified. Any QA program is wasted effort if the statistical control of a process has not been documented. The QA program to provide the "user of a product or a service the assurance that it meets defined standards of quality with a stated level of confidence" (Taylor, 1987). When the uncertainty of a process has been documented and these data meet the predetermined limits of the QA program, the process is said to be within statistical control.

Another type of diagram, often called a "fishbone diagram," is useful for breaking a process into the parts which directly contribute to the final quality of the result (Ishikawa, 1976). Figure 3 shows a fishbone diagram (nonexhaustive) that illustrates possible sources of uncertainty for a nitrate colorimetric process. Each main component contributing to uncertainty can further be broken down and indicated as additional branching within the main component. The manager can then make choices concerning further testing of a process component if the overall uncertainty of the process is not within acceptable limits.

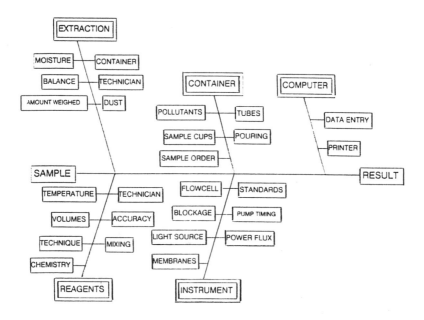

Figure 3.   "Wish Bone" Diagram Illustrating Sample Flow Through the Analytical Laboratory.

If the process is found to be within current limits, these diagrams can help the manager to decide where to focus attention in order to further reduce current limits. Additionally, these diagrams are useful for training new technicians since the diagrams point out the components of the process which contribute to uncertainty.

Statistical Measurements

The measurements needed to insure that the process is within acceptable limits can take many forms, some of which are discussed below. No subset of these measurements is best for all processes. The manager must choose the method or methods that are best for the particular process within a specific laboratory. The following list is not exhaustive, but provides a wide spectrum of possibilities.

In this context, the term "external standards" refers to standards that are prepared by some source outside of the laboratory. The National Institute of Standards and Technology, for example, specifies the mean measurement with some level of confidence about that mean. Thus, the external standard provides an unbiased estimate of the laboratory's ability to determine accurately a value for that measurement.

Accuracy is the agreement of the measured value from the laboratory and the expected value supplied by the source of the external standard. Use of the external standard is considered unbiased because the laboratory itself does not specify the expected value.

An internal standard is a reference material that has been prepared within the laboratory. While of great value to the QA program, such standards can be considered biased because the laboratory itself supplies both observed and expected values.

The manager should explore a mixture of both external and internal standards to insure accuracy. Acceptance of this mixture may be directly dependent on the needs of the client.

Most external standards are costly, and frequent use increases the operating costs of the laboratory. Internal standards are usually much less expensive. Thus, if a relationship between external and internal standards can be firmly established, frequent use of the internal standard with occasional use of the external standard can reduce costs while still providing acceptable QA.

Replicates of Unknowns

Repeated measurements of an unknown sample give an estimate of the precision of the process. Precision of the process is the level of agreement among independent measurements using the specified conditions of the process. Use of this QA technique can provide information on the stability of the process to reproduce the values within a certain level of uncertainty. It has the added advantage of generating such results directly from the client's samples, not from standards which may have a differing matrix from the client's samples. In addition, the material is supplied by the client, usually at no extra cost to the laboratory.

As with external and internal solutions, the manager must strike a balance between the number of replicates and the production of completed samples by the laboratory. When the process is in statistical control, replicates can be held to a minimum, as, for example, one duplicate in every 20 unknown samples. This number should increase if the process is working within about five times the value of the process lower limit of detection.

Spiked Unknowns

This method of determining quality measures the percentage of recovery of an internal standard when it is mixed with an unknown sample. While often used in QA programs, this method may introduce dissimilar matrix problems into the process. Additional errors can be introduced in the measurement and selection of the quantity and level of the spiking solution. Taylor (1989) advised that spiking should not be used for calibration and validation when other means are available. For this reason, spiking should not be frequently conducted on samples for normal agricultural testing. While this procedure may be of some benefit to the QA program, it can also introduce serious problems.

CONTROL CHARTS

Control charts can be the backbone of a good QA program. Data collected on a frequent basis from these charts can be combined to produce process audits by the manager and summary QA documents for the client. Some of the many types of control charts—those that can be most easily used by the technical staff—are presented here.

X charts can be used for both external and internal standards. Figure 4 shows the daily values for an internal standard of 10 mg NO_3-N/L with time used at the Extension Soil Testing Laboratory, University of Florida. Daily readings such as these demonstrate the time stability of the process. Drifting or out-of-control values can be detected at the work station and recalibration initiated before large numbers of unknown samples must be rerun. If the QA program is truly of highest concern within all parts of the laboratory, the goal will not be to produce acceptable or "within specification" results. Rather, the goal will be to produce what can be achieved in the given process.

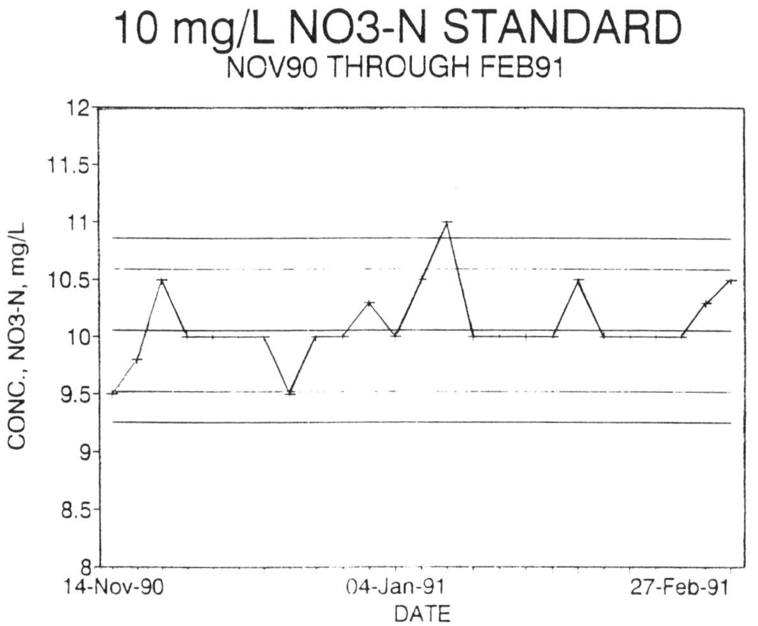
10 mg/L NO3-N STANDARD
NOV90 THROUGH FEB91
CONC., NO3-N, mg/L
12
11.5
11
10.5
10
9.5
9
8.5
8
14-Nov-90
04-Jan-91
27-Feb-91
DATE

Figure 4. *X Chart* Showing the Daily Values for an Internal Standard.

Figure 4 also shows both upper and lower warning limits (UWL and LWL) and control limits (UCL and LCL). These limits are found by calculating the standard deviation of a portion of the data (a minimum of 7 points is recommended) and multiplying by 2 for the WL and 3 for the CL. The UWL and LWL lines will include about 95% of the data, while the UCL and LCL lines will include 99%. Rules for deciding about the statistical control of the process are discussed later in this chapter.

A similar chart using the mean of four or more analyses, called the *X-bar chart*, can also be constructed. However, this chart does require many more additional analyses, and for this reason it is not frequently used in agricultural testing.

The *R chart*, alternately called a *range chart*, is a useful method of displaying the range (highest-lowest) of replicated samples. The data displayed in Figure 5 are from duplicate analyses. The mean R is calculated on a minimum of 8 values, but preferably with 15 values (Taylor, 1987). The UWL is calculated by multiplying the standard deviation of the data by 2.512, while the UCL is found by multiplying the standard deviation by 3.267.

The R chart will not reflect some process-oriented problems, such as deteriorating instrument calibration. For this reason, the R chart should be used in conjunction with an X chart. Since the R chart is developed by replicating unknown samples, this chart accurately indicates the precision of the process.

The R chart can be used with replicates of other than duplicate size (Taylor, 1987). The R chart should be used for replicate samples that are within a reasonable range of values. Replicated samples that are of widely differing reported values should be used with caution. Selection of the range must be based upon the process. A series of tests is needed for the selection.

Another option, the *range performance chart*, allows a wide range of readings (Figure 6). The chart does not reflect the precision of the system chronologically. This chart was made based upon data that were collected within four concentration groups. Calculations are the same as for the R chart, but mean R values and limits are plotted and a straight line drawn through respective points.

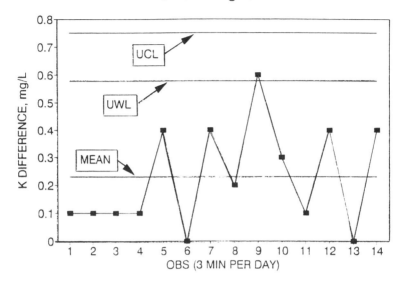

Figure 5. *R Chart* (or *Range Chart*) Displaying the Range (highest-lowest) of Replicated Samples.

Making Decisions with the Charts

Daily use of the charts can be the first line of defense against poor quality data. In using the charts, the technician should follow a few general rules:

1. If two measurements fall within the warning (UWL or LWL) and control lines (UCL or LCL), reject all data which lie between the most recent reading and the last control sample that is known to be in control (reading within the UWL and LWL). Take corrective action, and reread the identified samples.

2. If one measurement falls outside of the control lines (UCL or LCL), reject the data from the most recent sample and the last control sample which was known to be in control. This rule applies to the high NO_3-N reading which exceeded the UCL

K RANGE PERFORMANCE CHART
RANGE 0 TO 500 mg K/L

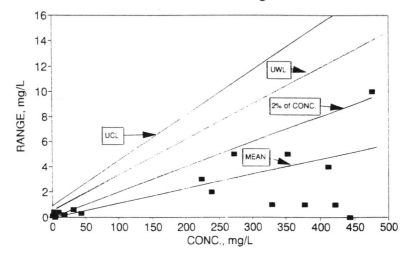

Figure 6. *Range Performance Chart* which allows for a
Wide Range of Readings to be Graphed.

(Figure 4). Samples completed since the last in-control standard
(10.5 mg NO_3-N/L) would be rejected. The process would be
recalibrated and all rejected samples retested.

3. Identify trends in the data (more than 7 points in rough
alignment) and take corrective action before control samples fall
outside the warning lines.

BLIND STUDIES

Two levels of blind studies are possible. The first level is
composed of one or more samples of known values (external
and/or internal standards). The technician is usually aware that
these samples will be used in the QA program. However, the
technician does not know the expected analytical results.

The second level, the so-called *double blind study,* is similar to the first system in every respect, except that the technician cannot distinguish the QA samples from other unknown samples. For example, the QA officer enters a set of external/internal samples into the laboratory under an assumed name. None of the laboratory personnel are aware of the nature of these samples and the set receives the same attention as unknown samples.

The double blind system is the preferred method of collecting data about the health of the QA process, since it removes any preconceived bias present with the first system. When the process is considered in statistical control, such studies (either blind or double blind) can be conducted weekly with good results. If the process is not in control, blind studies should not be conducted. Rather, efforts should be focused on establishing control.

SYSTEMS AUDITS

Systems audits are conducted to insure that the process is within specified QA objectives. The systems audit should review all of the statistical data generated from the process as well as instrument logs and in-laboratory time for analyses. The discussions should address those parts of the process which (1) are the most difficult, (2) have the most hand labor, or (3) are just unpopular. It is the technician's responsibility to detail real or perceived impediments to quality, while the manager must inspect the process for means to remove these impediments.

A systems audit may point out problems with which a technician cannot deal, either due to a lack of training or a lack of specific guidance. Training of both technical and management staff, or even a brief restatement of the goals, is often the solution to process problems.

DOCUMENTATION

The following parameters will establish the basic paperwork needed for a sound QA program. In all cases, documentation should be used to substantiate the QA program. Paperwork in and of itself is of little value unless it can be created easily, can be interpreted by those who must use it, and can be read by those summarizing the information contained therein. All documentation must be functional and not merely contribute to overhead costs.

1. Custody of Samples

The progress of samples through the laboratory must be accurately documented. Simple forms or computer-generated screens can be used to detail a clear path of sample receipt by the laboratory, showing any work performed within the laboratory and by whom, subsequent sample disposal, and information reported to the client.

Multiple copies of such paperwork should be strongly questioned. For example, it makes little sense for individual technicians to maintain copies. Rather, the central office can maintain the original and a working copy can accompany the samples through the laboratory. In most laboratory information management systems, custody records can be computerized.

2. Standard Operating Procedures (SOP) or Good Laboratory Procedures (GLP)

These detailed documents (SOP and GLP) are needed to describe the internal functions of the laboratory. They are easily written after the within-laboratory supplier/customer pathways and the fishbone charts are diagrammed. If the process is new to the laboratory, one might further delay until the diagrammed process has been found to be within statistical control. Then, the procedural description will include those techniques which can be used to produce a stable process.

Many formats for SOP and GLP exist. Both the United States Environmental Protection Agency (EPA) and the Food and Drug Administration (FDA) have specific guidelines for such documents. State agencies usually will also have specific requirements for acceptable SOP/GLP documentation.

3. Methodology References

All methodology reference material should be readily available. Many laboratory manuals containing SOP/GLP information also contain the appropriate references. The purpose of such documentation is to show the implementation of the procedure as well as any interferences, etc., which influence the process. Such a reference is useful both to technicians and managers to insure that the procedure is being accurately followed. Any changes in the reference methodology should also be documented to avoid confusion and misunderstanding in the future.

4. Instrument Handling and Maintenance Procedures

Detailed descriptions of proper instrument techniques should be available. A readily available instrument manual may suffice for this guide, but often the details of a specific process may not be addressed in the manufacturer's manual. For this reason, a supplemental guide can be quite helpful, especially if the guide is written with the involvement of the technical staff. A checklist format for daily operations may also help to preclude unsafe procedures or forgetfulness.

Maintenance or operating records on each instrument are of great value since the history of problems and cures is then readily available. An additional benefit of such records is that instrument operating costs and sample volume can be collected with little loss of time. To be accurate, the operating record should include any samples which were reanalyzed due to rejected QC standards by the technical staff. Such samples may not be counted in any other way, yet they directly affect the productivity of the laboratory.

Preventive maintenance can also be scheduled in such records to alert the operator of needed care. Early detection and replacement of failing parts can have a positive effect on instrument performance.

5. QA Expectations and Measured Performance

Each laboratory should record and maintain information regarding each process within the laboratory. A clear statement of the QA objectives for a process can be quite helpful for the client. Such a statement also sets the specifications within which the technical staff should work. The overall management goal must be to do the best job possible. Specifications should be used only as guidelines, not as a listing of what is tolerable.

The measured performance of each process demonstrates that the process is within statistical control. Ideally, the performance should be better than the specifications, except where the specifications force work at or near the limits of some part of the process, such as the lower limit of detection of an instrument.

SUMMARY

While a QA program may be viewed as a drain on laboratory resources, such an assessment is not valid wherever an efficient QA program has been adopted. In the soil testing laboratory, a workable QA program has three components:

1. QA is always cost effective when used as a management tool to create a process which is in statistical control.

2. The trained technician is the best QA guarantee.

3. The laboratory manager modifies the process while the technician works within the process.

Both the technician and the manager must work together to remove all impediments to quality.

REFERENCES

1. Ishikawa, K. 1976. Guide to quality control. Unipub, Asian Productivity Organization, Murray Hill Station, NY.

2. Deming, W. E. 1986. Out of the crisis. Massachusetts Institute of Techhology, Center for Advanced Engineering Study, Cambridge, MA.

3. Lewis, J. K. 1987. Quality assurance of chemical measurements. Lewis Publishers, Clelsea, MI.

DETERMINATION OF SOIL WATER pH

1.   PRINCIPLE OF THE METHOD

1.1   This procedure is used to determine the pH of a soil in a water suspension. pH is defined as "the negative logarithm to the base 10 of the H-ion concentration, or the logarithm of the reciprocal of the H-ion concentration in the soil solution." Since the pH is logarithmic, the H-ion concentration in solution increases ten times when the pH is lowered one unit.

2.   RANGE AND SENSITIVITY

2.1   Commercially available standard pH meters can be used to measure the pH in water of soils between pH 3.5 and pH 9.0.

2.2   The sensitivity will depend on the instrument. In routine soil testing, it is only necessary to read the pH to the 0.1 unit.

3.   INTERFERENCES

3.1   Hydrogen ions may be displaced from the exchange sites, and the presence of other ions causes additional H-ions to form in solution. This interference results in a lower pH (see 12.5).

3.2   Carbon dioxide (CO_2) from the atmosphere or soil air dissolves in water, forming carbonic acid (H_2CO_3), which can lower the pH markedly. Only in soils that have a pH considerably above 7.0, i.e., a very low H-ion concentration, does the CO_2 concentration of the air have an appreciable effect on the pH measurement.

4.   PRECISION AND ACCURACY

4.1   Random variation of 0.1 to 0.2 pH unit is allowable in replicate determinations, and this variation can be expected from one laboratory to another.

4.2   Be careful to prevent scratching, since a scratched glass electrode will give erratic values. Likewise, reference electrodes which restrict the flow of the filling solution might cause unstable readings.

4.3   Dehydrated electrodes give erratic readings. Follow the electrode manufacturer's instructions in keeping the electrodes hydrated.

5. APPARATUS

5.1   No. 10 (2-mm opening) sieve.

5.2   4.25-cm^3 volumetric scoop for soils to be used for SMP buffer pH determination (see page 56). 10-cm^3 volumetric scoop for soils to be used for Adams-Evans buffer pH determination (see page 51).

5.3   50-mL cup (glass, plastic, or waxed paper of similar size).

5.4   Pipettes, 5-mL and 10-mL capacity.

5.5   Stirring apparatus (mechanical shaker, stirrer, or glass rod).

5.6   pH meter, line or battery operated with reproducibility to at least 0.05 pH unit, and a glass electrode paired with a calomel reference electrode.

5.7   Glassware and dispensing apparatus for the preparation and dispensing of buffer solutions.

5.8   Analytical balance.

6. REAGENTS

6.1   pH 7.0 Buffer Solution - Dissolve 3.3910 g citric acid ($C_6H_8O_7$) and 23.3844 g disodium phosphate ($Na_2HPO_4 \cdot 12H_2O$) in pure water and dilute to 1 liter (a commercially available buffer is acceptable).

6.2   pH 4.0 Buffer Solution - Dissolve 11.8060 g citric acid ($C_6H_8 O_7$) and 10.9468 g disodium phosphate ($Na_2HPO_4 \cdot 12H_2O$) in pure water and dilute to 1 liter (a commercially available buffer is acceptable).

7. PROCEDURE

7.1   Weigh 5 g, or scoop 4.25 cm^3 for a SMP buffer (see page), or scoop 10 cm^3 for an Adams-Evans buffer (see page) <10-mesh (2-mm) soil into a cup (see 5.3). Pipette 5 mL or 10 mL (for 4.25-cm^3 or 10- cm^3 soil, respectively,) into the cup and stir for 5 seconds. Let it stand for 10 minutes. Calibrate the pH meter according to the instructions supplied with the specific meter. Stir the soil and water slurry (see 5.5). Lower the electrodes into the soil-water slurry so that the electrode tips are at the soil-water interface. While stirring the soil water slurry, read the pH and record to the nearest tenth of a pH unit.

7.2 Save the sample for the determination of the buffer if the soil water pH is less than 6.0.

8. CALIBRATION AND STANDARDS

8.1 To calibrate the pH meter, use prepared (see 6.1, 6.2) or commercially available buffer solutions of pH 7.0 and pH 4.0, according to the instruction manual.

9. CALCULATION

9.1 The result is reported as water pH (pH_W).

10. EFFECTS OF STORAGE

10.1 Air-dry soils may be stored several months in closed containers without affecting the pH_W measurement.

10.2 If the pH meter and electrodes are not to be used for extended periods of time, the instructions for storage published by the instrument manufacturer should be followed.

11. INTERPRETATION - see 12.1, 12.2, 12.3, and 12.4.

12. REFERENCES

12.1. Schofield, R. K., and A. W. Taylor. 1955. The measurement of soil pH. Soil Sci. Soc. Amer. Proc. 19:164-167.

12.2. Coleman, N. T., and G. W. Thomas. 1967. The basic chemistry of soil acidity, pp. 1-41. IN: R. W. Pearson and F. Adams (eds.), Soil Acidity and Liming. American Society of Agronomy, Madison, WI.

12.3. Peech, M. 1965. Hydrogen-ion activity, pp. 914-926. IN: C. A. Black (ed.), Methods of Soil Analysis, Part 2, Agronomy No. 9. American Society of Agronomy, Madison, WI.

12.4. McLean, E. O. 1973. Testing soils for pH and lime requirement, pp. 78-95. IN: L. M. Walsh and J. D. Beaton (eds.), Soil Testing and Plant Analysis, rev. ed. Soil Science Society of America, Madison, WI.

12.5. Coleman, N. T., and G. W. Thomas. 1964. Buffer curves of acid clays as affected by the presence of ferric iron and aluminum. Soil Sci. Soc. Amer. Proc. 28:187-190.

DETERMINATION OF SOIL pH IN 0.01M CaCl$_2$

1. PRINCIPLE OF THE METHOD

1.1   This method estimates the activity of H-ions in a soil suspension in the presence of 0.01M CaCl$_2$ to approximate a constant ionic strength for all soils, regardless of past management, mineralogical composition and natural fertility level.

1.2   The use of 0.01M CaCl$_2$ in soil pH measurement was proposed by Schofield and Taylor (see 12.1). Peach (see 12.2) summarized the advantages of using 0.01M CaCl$_2$ for measuring soil pH. The merits of determining soil pH in a constant salt level are also discussed by McLean (see 12.3), Woodruff (see 12.4) and Conyers and Darey (see 12.5).

2. RANGE AND SENSITIVITY

2.1   Commercially available standard pH meters can be used to measure the pH in 0.01M CaCl$_2$ of soils that range from pH 2.5 to pH 8.0, which would include most soils.

2.2   The sensitivity will depend on the instrument. In routine soil testing, it is only necessary to read the pH to the 0.1 unit.

2.3   The pH in 0.01M CaCl$_2$ may be estimated with brom cresol purple (see 12.4).

3. INTERFERENCES

3.1   The main advantage of the measurement of soil pH in 0.01M CaCl$_2$ is the tendency to eliminate interferences from suspension effects and from variable salt contents, such as fertilizer residues.

4. PRECISION AND ACCURACY

4.1   Temperate region soil pH values in 0.01M CaCl2 are lower in magnitude (higher H-ion concentration) and less variable than those made in water, due to the release of H-ions from exchange sites by calcium ions.

4.2   See 4.1, 4.2 and 4.3 on page 1.

5. APPARATUS

5.1   No. 10 (2-mm opening) sieve.

5.2   4.25-cm^3 scoop, volumetric.

5.3   50 mL-cup, (glass, plastic, or waxed paper of similar size).

5.4   Pipette, 5-mL capacity.

5.5   Stirring apparatus (mechanical shaker, stirrer or glass rod).

5.6   pH meter, line or battery operated, with reproducibility to at least 0.05 pH unit, and a glass electrode paired with a calomel reference electrode.

5.7   Glassware and dispensing apparatus for the preparation and dispensing of 0.01M CaCl$_2$ and buffer solutions.

5.8   Dropping bottle, 30- or 60-mL capacity (see alternate procedure 7.2 below).

5.9   Analytical balance.

6. REAGENTS

6.1   <u>0.01M Calcium Chloride</u> - Weigh 1.47 g calcium chloride dihydrate (CaCl$_2 \cdot$ 2H$_2$O) into a 1-liter volumetric flask and dilute to the mark with pure water.

6.2   <u>pH 7.0 Buffer Solution</u> - Dissolve 3.391 g citric acid (C$_6$H$_8$O$_7$) and 23.3844 g disodium phosphate (Na$_2$HPO$_4 \cdot$ 12H$_2$O) in pure water and dilute to 1 liter (a commercially available buffer is acceptable).

6.3   <u>pH 4.0 Buffer Solution</u> - Dissolve 11.806 g citric acid (C$_6$H$_8$O$_7$) and 10.9468 g disodium phosphate (Na$_2$HPO$_4 \cdot$ 12H$_2$O) in pure water and dilute to 1 liter (a commercially available buffer is acceptable).

6.4   <u>1.0 M Calcium Chloride</u> - Weigh 147 g calcium chloride dihydrate (CaCl$_2 \cdot$ 2H$_2$O) into a 1-liter volumetric flask and dilute to the mark with pure water.

7. PROCEDURE

7.1   Weigh 5 g air-dry, or scoop 4.25 cm^3 <10-mesh (2-mm) soil into a 50-mL cup (see 5.3). Pipette 5 mL 0.01M CaCl$_2$ solution (see 6.1) into the cup and stir for 30 minutes on a mechanical stirrer or shaker (or stir periodically with a glass rod for a period of 30 minutes). Calibrate the pH meter according to the instructions supplied with the specific meter. Stir the soil and 0.01M $CaCl_2$ slurry. Lower the electrodes into the soil-0.01M $CaCl_2$ slurry so that the electrode tips are at the soil-water interface. While stirring the soil water slurry, read the pH and record to the nearest tenth of a pH unit.

7.2   For laboratories desiring both soil pH in water and 0.01M $CaCl_2$, 5 mL pure water can be substituted for the 5 mL 0.01M $CaCl_2$, as given in 7.1. After the water pH is determined, add one drop of 1M $CaCl_2$ (see 6.4) to the soil-water suspension, stir or shake for 30 minutes and then read the pH of the suspension and designate as pH_{Ca}.

8. CALIBRATION AND STANDARDS

8.1   The pH meter is calibrated using prepared (see 6.1, 6.2) or commercially available buffer solutions of pH 7.0 and pH 4.0, according to the instruction manual.

9. CALCULATION

9.1   The result is reported as water pH (pH_s).

10. EFFECTS OF STORAGE

10.1   Air-dry soils may be stored several months in closed containers without affecting the pH_s measurement.

10.2   If the pH meter and electrodes are not to be used for extended periods of time, the instructions for storage published by the instrument manufacturer should be followed.

11. INTERPRETATION - See 12.5 or 12.6.

12. REFERENCES

12.1.   Schofield, R. K., and A. W. Taylor. 1955. The measurement of soil pH. Soil Sc. Soc. Am. Proc. 19:164-167.

12.2.   Peech, M. 1965. Hydrogen-ion activity, pp. 914-926. IN: C. A. Black (ed.), Methods of Soil Analysis, Part 2, Agronomy No. 9. American Society of Agronomy, Madison, WI.

12.3.   McLean, E. O. 1973. Testing soils for pH and lime requirement, pp. 78-95. IN: L. M. Walsh and J. D. Beaton (eds.), Soil Testing and Plant Analysis, rev. ed. Soil Science Society of America, Madison, WI.

12.4.   Woodruff, C. M. 1961. Brom cresol purple as an indicator of
        soil pH. Soil Sci. 91:272.

12.5   Conyers, M. K., and B. G. Darey. 1988. Observations in some
        routine methods for soil pH determination. Soil Sci. 145: 29-
        36.

12.6   Graham, E. R. 1959. An explanation of theory and methods of
        soil testing. Missouri Agri. Expt. Sta. Bull. 734.

12.7   Woodruff, C. M. 1967. Crop response to lime in the mid-
        western United States, pp. 207-227. IN: R. W. Pearson and
        F. Adams (eds.), Soil Acidity and Liming. American Society of
        Agronomy, Madison, WI.

DETERMINATION OF SOIL-PASTE pH AND CONDUCTIVITY OF SATURATION EXTRACT

1. PRINCIPLE OF THE METHOD

1.1   Plants growing in salt-affected soils are affected by the osmotic pressure of the soil solution. At field capacity, coarse-textured soils hold less water than do fine-textured soils and thus have higher osmotic pressures in soil solutions at the same salt concentration on a soil-weight basis. Salt concentration in the saturated paste extract is related to the water-holding capacity of soils at field capacity and thus will alleviate the problem of salt concentration on a soil-weight basis.

1.2   Using saturated paste to measure the salt content of soils was first proposed by Schofield in a report to the US National Resources Planning Board (see 12.1) and was subsequently adopted by the US salinity laboratory staff (see 12.2). Measuring the pH of soil paste was suggested by the US salinity laboratory staff (see 12.2).

2. RANGE AND SENSITIVITY

2.1   The method is adapted to a wide range of saturated paste conductivities and pH values. The extract can be diluted if its conductivity is outside the range of the meter used.

3. INTERFERENCES

3.1   If the water content is higher or lower than the saturation point, the conductivities will be affected.

3.2   The electrodes should be well-platinized for reproducible results.

3.3   The electrical conductivity increases with an increase in temperature.

3.4   For pH measurements, the temperature of the buffer solution and the soil paste should be about the same.

4. PRECISION AND ACCURACY

4.1   Conductivity values of less than one should be reported to two decimal places and values more than one to three significant figures (see 12.3). For soil-paste pH, values should be reported to the nearest tenth.

4.2   See 4.1, 4.2 and 4.3, page 1.

5. APPARATUS

5.1   No. 10 (2-mm opening) sieve.

5.2   150-mL beaker.

5.3   250-mL vacuum flask.

5.4   Vacuum pump.

5.5   Spatulas.

5.6   Filter paper, Whatman No. 5 or equivalent.

5.7   Filter funnel stand.

5.8   Buchner funnel.

5.9   Conductivity bridge with 0 to 1 million ohms capacity.

5.10   Conductivity cell, pipette type.

5.11   pH meter, line or battery operated, with reproductibility to at least 0.05 pH unit, and a glass electrode paired with a calomel reference electrode.

6. REAGENTS

6.1   0.01N Potassium Chloride - Dissolve 0.7456 g potassium chloride (KCl) in pure water, and add pure water to make 1 liter at 25° C. This standard solution has an electrical conductivity of 0.0014118 mho per cm at 25° C.

6.2   Buffer Solution - pH 4.0 and pH 7.0 buffers for standardization of the pH meter (see page 2).

7. PROCEDURE

7.1   First add a small amount of pure water to a beaker. Then fill the beaker approximately 2/3 full of soil. Add pure water to the soil in the beaker while stirring with a spatula. At saturation, the soil paste glistens as it reflects light, flows slightly when the container is tipped, and the paste slides freely and cleanly off the spatula for all soils but those with a high clay content. After mixing, the samples should be allowed to stand for an hour or more, and then the criteria for saturation should be rechecked. Free water should not collect on the soil surface nor should the paste stiffen markedly or lose its glistening appearance on standing. If the paste is too wet, additional dry soil may be added. For clay soils, the water should be added with a minimum of stirring. Peat and muck soils require an overnight wetting period to obtain a definite end point for the saturation paste.

(Note: The sodium, calcium, and magnesium content of the extract can be determined to evaluate the sodium hazard, as in 12.2.)

7.2   A pH reading of the paste is recorded, then the saturated soil paste is transferred to the Buchner funnel with a filter paper in place. A vacuum is applied and the extract is collected in a 250-mL vacuum flask. The temperature of the extract is measured. The conductivity cell is rinsed and filled with the extract. The temperature compensation dial, if available, is set at the extract temperature, or the reading is corrected to a temperature of $25°C$ (see 12.2). The electrical resistance of the extract is read.

8. CALIBRATION AND STANDARD

8.1   Measure the resistance of the 0.01N potassium chloride (KCl) solution and correct the reading to $25°C$ (see 12.2). This value is used in calculating the conductivity of the saturation extract as explained under 9.1. If the temperature of the 0.01N KCl solution and that of the saturation extract are the same, no temperature correction is necessary.

9. CALCULATIONS

9.1   Electrical conductivity (EC) of the extract is calculated as follows:

$$EC, \text{ mhos per cm at } 25° C = 0.0014118 \times R_{std}/R_{ext}$$

where the value of 0.0014118 is the electrical conductivity of the standard 0.01N potassium chloride (KCl) solution in mho per cm at $25°C$ and R_{std} and R_{ext} refer to electrical resistance of the standard (0.01N KCl) solution and extract, respectively. Multiply the results by 1000 to obtain mmho per cm at $25°C$. Report conductivity values of less than one to two decimal places, and values of more than one to three significant figures.

10. EFFECTS OF STORAGE

10.1   Air-dry soils may be stored without any appreciable effect on pH and conductivity.

11. INTERPRETATION

11.1   Crops are different in their sensitivity to salt content. For the interpretation of results, see USDA Agric. Handbook No. 60 (12.2).

12. REFERENCES

12.1   United States National Resources Planning Board. 1942. The Pecos River Joint Investigation: Reports of participating agencies. US Government Printing Office, Washington DC.

12.2   Salinity laboratory staff. 1954. Diagnosis and improvement of saline and alkaline soils. Agriculture Handbook No. 60, USDA. US Government Printing Office, Washington, DC.

12.3   Bower, C. A., and L. U. Wilcox. 1965. Soluble salts, pp. 939-949. IN: C. A. Black (ed.), Methods of Soil Analysis, Part 2, Agronomy No. 9. American Society of Agronomy, Madison, WI.

DETERMINATION OF SPECIFIC CONDUCTANCE IN SUPERNATANT 1:2 SOIL:WATER SOLUTION

1. PRINCIPLE OF THE METHOD

1.1   Although specific conductance measurements in saline soils are principally carried out on a soil-paste extract, research workers in the humid soil region make extensive use of a 1:2 soil:water extraction of greenhouse soils. Specific conductance values in the 1:2 extract have been observed not to be comparable with those in the saturation extract. However, Jackson (see 12.1) concluded that specific conductance ranges of the widely contrasting alkaline and humid regions are quite similar.

1.2   The specific conductance method described is based on the experience of the Agronomic Division, North Carolina Department of Agriculture. All measurements were made on greenhouse soils and on field problem soils. The 1:2 soil:water ratio in the procedure is based on a soil volume rather than on a soil weight basis. This avoids the need for further dilution of low bulk densities for Histosols. Guidelines for restoring fields flooded by salt water are included.

2. RANGE AND SENSITIVITY

2.1   The method is adapted to a wide range of salt concentrations, depending on the instrument. The range can be extended by suitable dilution of the extract.

3. INTERFERENCES

3.1   Specific conductance increases with increasing temperature; hence, compensation of temperature differences from the calibrated standard is required.

3.2   For reproducible results, clean and well-platinized electrodes are essential.

4. PRECISION AND ACCURACY

4.1   Conductivity values of less than one (<1.0) should be reported to two decimal places and values more than one (>1.0) to three significant figures (see 12.3). For soil-paste pH, values should be reported to the nearest tenth.

4.2   See 4.1, 4.2 and 4.3, page 1.

5. APPARATUS

5.1   No. 10 (2-mm opening) sieve.

5.2   50-60 mL cup, (glass, plastic or waxed cup of similar size).

5.3   10-cm^3 scoop, volumetric (see 12.3)

5.4   20-mL pipette.

5.5   Conductivity bridge with 0 to 1 million ohms capacity.

5.6   Conductivity cell, pipette type, 2- to 3-mL capacity.

5.7   Thermometer, 1-100° C.

6. REAGENTS

6.1   <u>0.01N Potassium Chloride</u> - Dissolve 0.7456 g potassium chloride (KCl) in pure water. Make up to 1 liter with pure water.

7. CALIBRATION AND STANDARDS

7.1   To determine the cell constant (θ), use the 0.01N KCl (see 6.1) solution at 25° C, which will will have a specific conductance (SC) of 0.0014118 mho per cm.

7.2   The cell constant (θ) for any commercially available conductivity cell can be calculated, according to Willard, Merritt and Dean (see 12.4, page 720), by the relationship:

$$K = (1/R)\ (d/A) = \theta/R$$

where K = specific conductance, A = electrode area, d = plate spacement, and R = resistance in ohms per cm. In the case of 0.01N KCl (see 7.1), the cell constant (θ) = 0.0014118 (in mho per cm) x R (in ohms per cm). R = 708.32 ohm if the cell has electrodes 1 cm^2 in an area spaced 1 cm apart. (Notice that mhos = 1/ohm).

7.3   Some conductivity instruments read in specific conductance (SC) are expressed in mhos x 10^{-5} as well as resistance (ohms). Before accepting the mhos x 10^{-5} dial readings, the cell constant should be determined and the mhos x 10^{-5} dial readings substantiated as being correct for the cell constant used.

8. PROCEDURE

8.1   Scoop 10-cm^3 (see 5.3) of 2-mm sieved soil into a beaker (see 5.5), add 20 mL pure water, and stir thoroughly. Allow the suspension to settle for at least 30 minutes or long enough for the solids to settle.

8.2   Draw the supernatant into the conductivity pipette to slightly above the constricted part of the pipette. Avoid drawing the liquid into the rubber bulb. If this occurs, rinse the bulb before continuing with the next sample.

9. CALCULATIONS

9.1   Specific conductance (SC) of the soil extract is calculated as follows:

$$SC, \text{ mhos per cm at } 25^\circ C = (0.0014118 \times Rstd)/Rext$$

where the value of 0.0014118 is the specific conductance of the standard 0.01N KCl solution in mho per cm at 25° C and Rstd and Rext refer to resistance in ohms of the standard (0.01N KCl) solution (see 6.1) and extract (see 8.1), respectively. Multiply the results by 1000 to obtain mmho per cm at 25° C. Report specific conductance values in mmho per cm.

9.2   Alternate method of calculation: After the cell constant (θ) has been determined (see 7.2), the specific conductance of the soil extract can be obtained from the following relationship:

$$SC, \text{ mhos per cm at } 25^\circ C = \theta/R$$

where θ = determined cell constant and R = resistance in ohms of the soil extract.

10. INTERPRETATION

10.1   Results with various soils and crops using the 1:2 soil:water ratio extraction have been reported by Dunkle and Merkle (see 12.6) and Merkle and Dunkle (see 12.7). Jackson (see 12.1) has summarized the relationships of specific conductance in 1:2 soil:water extract (observed) to that in the saturation extract (calculated) for a silt loam soil at 40% saturation, and a clay loam high in organic matter soil at 100% saturation. The conductance ratios of the 1:2 saturation extract values of the 40% and 100% saturated soils were 0.2 and 0.5, respectively.

10.2 Using the 0.2 ratio values in relation to the Scofield salinity scale (see 12.5), together with published (see also 12.8) and local experience (Agronomic Division, North Carolina Department of Agriculture), a general guide to plant effects associated with different ranges of specific conductance measured in a 1:2 soil:water ratio by volume is as follows:

mmho/cm at 25° C	Effects
<0.40	Non saline: salinity effects mostly negligible, excepting possibly bean and carrot.
0.40-0.80	Very slightly saline: yields of very salt sensitive crops, such as flax, clover (alsike, red), carrot, onion, bell pepper, lettuce and sweet potato, may be reduced by 25% to 50%.
0.81-1.20	Moderately saline: yield of salt-sensitive crops restricted. Seedlings may be injured. Satisfactory for well drained greenhouse soils. Crop yields reduced by 25% to 50% may include broccoli and potato plus the other plants above.
1.21-1.60	Saline: crops tolerant include cotton, alfalfa, cereals, grain sorghums, sugar beet, bermuda grass, tall wheat grass and Harding grass. Salinity higher than desirable for greenhouse soils.
1.61-3.20	Strongly saline: only salt-tolerant crops yield satisfactory. For greenhouse crops, leach soil with enough water so that 2-4 quarts pass through each square foot of bench area or one pint of water per 6 inch pot; repeat after about 1 hour. Repeat again if readings are still in the high range.
>3.2	Very strongly saline: only salt-tolerant grasses, herbaceous plants, certain shrubs and trees will grow.

11. GUIDELINES FOR RESTORING FIELDS FLOODED BY SALT WATER

11.1 Agricultural extension specialists at NC State University and agronomists of the Agronomic Division, NC Department of Agriculture, suggest that the following procedures be used when cropland is flooded by salt water or salt spray is blown inland by hurricane-force winds.

11.2   Plants growing on salt-water flooded soil exhibit the greatest damage when soil moisture is limiting growth. When the soil is relatively dry, the salt concentration of the soil solution around the plant roots is the highest and prevents uptake of moisture. On the other hand, salt water may wash across fields doing little or no damage if the soil has been previously saturated by rain or fresh-water floods.

11.3   Treatment for returning land to a productive level is based on the salt content from properly collected samples of the suspected salt-damaged area. Collect core samples to a depth of 18 inches (7 cm) from each field. Divide cores into four parts as follows: (a) 0-2 inches; (b) 2-6 inches; (c) 6-12 inches and (d) 12-18 inches. Salt concentration is determined on a 1:2 soil:water sample and the results expressed in mmho/cm.

11.4   The effects and reclamation are as follows:

mmho/cm	Effects and reclamation
<0.4	Most crops will grow quite well; no injury should be expected.
0.4-0.8	Fairly safe for most crops; however, a long dry spell may draw salts up near surface and damage plants.
0.81-1.2	Only salt-tolerant crops as listed below will grow. Reclamation for other crops as suggested below.
>1.2	Few crops will survive; reclamation necessary.

11.5   Crop Tolerance to Salt:

   11.51   Tolerant: Rape, kale, cotton, barley, tall fescue, garden pea, Rhodes grass and bermuda. Turf grasses and ornamentals, i.e., zoysia, St. Augustine, American beachgrass, sea oat, English ivy, dwarf yaupon, several species of yucca, dwarf natal-plum, sea grape, Japanese privet, common oleander, and wax myrtle.

   11.52   Moderately Tolerant: Fig, grape, wheat, oat, rye, sunflower, corn, ryegrass, alfalfa, sweet clover, sudan grass, birdsfoot trefoil, orchardgrass, carrot, lettuce, onion and tomato.

11.53   Sensitive: Pear, peach, apple, plum, vetch, field beans, green beans, red clover, white clover, alsike clover, ladino clover, cabbage, potato and many others.

11.6   Reclamation

11.61   Procedure: Apply calcium sulfate (landplaster, gypsum) to the fields if mmho/cm is above 1.00 in the top 6 inches of the soil. The application rate is as follows: (a) less than 2% organic matter, 2,000 lbs/a; (b) 2% to 5% organic matter, 3,000 lbs/a; and (c) above 5% organic matter, 4,000 lbs/a.

11.7   Resample in 3 to 6 months to determine the progress of treatments. Since calcium sulfate contributes to the specific conductance, it is essential to determine calcium and sodium in the extract also.

12. REFERENCES

12.1   Jackson, M. L. 1958. Soil Chemical Analysis. Prentice-Hall, Inc., Englewood Cliffs, NJ.

12.2   Bower, C. A., and L. U. Wilcox. 1965. Soluble salts, pp. 933-951. IN: C. A. Black (ed.), Methods of Soil Analysis, Part 2, Agronomy No. 9. American Society of Agronomy, Madison, WI.

12.3   Mehlich, A. 1973. Uniformity of soil test results as influenced by volume weight. Comm. Soil Sci. Plant Anal. 4:475-486.

12.4   Willard, H. H., L. L. Merritt, Jr., and J. A. Dean. 1968. Instrumental Methods of Analysis, 4th ed. D. Van Nostrand Co., Inc., Princeton, NJ.

12.5   Salinity Laboratory Staff. 1954. Diagnosis and improvement of saline and alkali soils. USDA. Agricultural Handbook No. 60. US Government Printing Office, Washington, DC.

12.6   Dunkle, E. C., and F. G. Merkle. 1944. The conductivity of soil extraction in relation to germination and growth of certain plants. Soil Sci. Soc. Am. Proc. 8:185-188.

12.7   Merkle, F. G., and E. C. Dunkle. 1944. The soluble salt content of greenhouse soils as a diagnostic aid. J. Am. Soc. Agron. 36:10-19.

12.8 Reisenauer, H. M. 1978. Soil and plant tissue testing in California, Bull. 1987 (rev. ed.). Div. Agr. Sci., University of California, Berkeley, CA.

DETERMINATION OF SOIL BUFFER pH BY THE ADAMS-EVANS LIME BUFFER

1. PRINCIPLE OF THE METHOD

1.1   This procedure describes the determination of the lime requirement of a soil by the Adams-Evans buffer method (see 12.1). The method was developed for non-montmorillonitic, low organic matter soils where the amounts of lime needed are small and the possibility of damage from overliming exists. The lime requirement of an acid soil is defined by this procedure as the amount of lime or other base required to change an acid condition to a less acid condition (a maximum pH_w of 6.5).

1.2   The Adams-Evans lime requirement method is based on separate measures of soil pH determined in water and buffer pH. Soil pH is used as a measure of acid saturation of the soil, designated "H-sat$_1$" below (see 12.1), according to the following:

$$\text{Measured soil pH} = 7.79 - 5.55\,(\text{H-sat}_1) + 2.27\,(\text{H-sat}_1)^2$$

where H-saturation is expressed as a fraction of CEC. Buffer pH is used as a measure of soil acids, designated "soil H" below (see 12.1 and 12.2), according to the equation

$$\text{Soil H} = 8(8.00 - \text{buffer pH})$$

for a 10-g soil sample in 10 mL water +10 mL buffer where "soil H" is in meq/100 g of soil. A pH change of 0.01 in 20 mL of solution (10 mL of water + 10 mL buffer) is caused by 0.008 meq of acid at a pH level between 7 and 8. CEC is calculated by using "H-sat$_1$" and "soil H" according to the equation:

$$\text{CEC} = \text{Soil H}/\,\text{H-sat}_1$$

The desired soil pH (not to exceed 6.5) is expressed in terms of acid saturation (designated "H-sat$_2$" below), according to the following:

$$\text{Desired soil pH} = 7.79 - 5.55\,(\text{H-sat}_2) + 2.27\,(\text{H-sat}_1)^2$$

2. RANGE AND SENSITIVITY

2.1   The Adams-Evans buffer method is very reliable for soils with relatively small amounts of exchangeable acidity (max. = 8 meq/100 g). The procedure provides a fairly high degree of accuracy for estimating lime requirements to reach pH 6.5 or less.

2.2   Sensitivity for the lime requirement determination is within 500 lb/A of lime.

3. INTERFERENCES

3.1   No significant interferences.

4. SENSITIVITY

4.1   A sensitivity of 0.01 in pH units of the buffer-soil slurry is needed.

5. APPARATUS

5.1   No. 10 (2-mm opening) sieve.

5.2   10-cm^3 scoop, volumetric.

5.3   50-mL cup, glass, plastic or waxed paper.

5.4   Pipette, 10-mL capacity.

5.5   Mechanical shaker (180 oscillations per minute) or stirrer.

5.6   pH meter, line or battery operated, with reproducibility to at least 0.01 pH unit, and a glass electrode paired with a calomel reference electrode.

5.7   Glassware and dispensing apparatus for the preparation and dispensing of Adams-Evans buffer.

5.8   Analytical balance.

6. REAGENTS

6.1   Adams-Evans Lime Buffer Solution - Dissolve 74 g potassium chloride (KCl) in 500 mL pure water. Add 10.5 g potassium hydroxide (KOH) and stir to bring into solution. Add 20 g p-nitrophenol (HO $\cdot$ C$_6$H$_4$ $\cdot$ NO$_2$) and continue to stir. Add 15 g boric acid (H$_3$BO$_3$). Stir and heat, if necessary, to bring into solution. Dilute to 1 liter with pure water when cool.

7. PROCEDURE

7.1 Scoop 10-cm^3 air-dry, <10-mesh (2-mm) soil into a 50-mL cup. Add 10 mL pure water and mix for 5 seconds. Wait for 10 minutes and read the soil pH. Only on samples with pH_w less than 6.4, add 10 mL Adams-Evans Buffer Solution (see 6.1) to the cup. Shake 10 minutes (see 5.5) or stir intermittently for 10 minutes. Let stand for 30 minutes. Read the soil-buffer pH on a standardized pH meter (see 8.1). Stir the soil suspension just prior to reading the pH. Read the pH to the nearest 0.01 pH unit.

8. CALIBRATION AND STANDARDS

8.1 The pH meter is adjusted to read pH 8.00 in an equal volume Adams-Evans Buffer (see 6.1) and pure water.

9. CALCULATION

9.1 The Adams-Evans buffer method assumes that agricultural-grade limestone is about 2/3 effective in neutralizing acidity up to a soil pH of about 6.5, and allows for this by using a correction factor of 1.5. Thus, the lime requirement is the product of the following equation (see 1.2):

(Soil H)/H-sat_1 x (H-sat_1 - H-sat_2) x 1.5

or for 10-g soil in 10 mL water + 10 mL buffer, it is

$CaCO_3$ (ton/A) = 8 [(8.00 - buffer pH)/ H-sat_1] x (H-sat_1 - H-sat_2) x 1.5

10. EFFECTS OF STORAGE

10.1 Air-dry soils may be stored several months in closed containers without affecting the pH_{Adams} measurement.

10.2 If the pH meter and electrodes are not to be used for extended periods of time, the storage instructions published by the instrument manufacturer should be followed.

11. INTERPRETATION

11.1 The Adams-Evans buffer method was developed for soils that have a maximum soil H content of 8.00 meq/100 g, and which have H-sat_1 of 1.00 at about pH 4.5. However, it can be used with soils with more H by adding less than 10 g soil to 10 mL water + 10 mL buffer and multiplying by the appropriate dilution factor. It also can be used with soils that have pH values below 4.5 when H-sat$_1$ is 1.00 by changing the intercept of the pH equation by the appropriate amount (see 1.2). For example, a soil that has a pH of 4.0 when H-sat is 1.00 has the following relationship between pH and H-saturation:

$$\text{Soil pH} = 7.29 - 5.55 \,(\text{H-sat}) + 2.27 \,(\text{H-sat})^2$$

11.2   The lime requirement for low CEC soils (and with a pH of about 4.5 when H-saturated) can be determined in the following table. This table is based on the pH:Adams-Evans buffer values.

11.3   Lime Requirement Table - Limestone (Ag-Ground, TNP = in 1000 pounds per acre), to raise soil pH to 6.5 to a depth of 6 2/3 inches.

Buffer pH	Soil water pH									
	6.2	6.0	5.8	5.6	5.4	5.2	5.0	4.8	4.6	4.4
					1000 lb/A					
7.95	0	0	0	0	2	2	2	2	2	2
7.90	0	0	0	0	2	2	2	2	2	2
7.85	0	0	2	2	2	2	2	2	2	2
7.80	0	1	2	2	2	2	2	2	2	2
7.75	0	1	2	2	2	2	2	2	2	2
7.70	1	1	2	2	2	2	2	2	2	
7.65	1	1	2	2	2	2	3	3	3	3
7.60	1	2	2	2	2	3	3	3	3	4
7.55	1	2	2	2	3	3	3	4	4	4
7.50	1	2	3	3	3	3	4	4	4	5
7.45	2	2	3	3	3	4	4	4	5	5
7.40	2	2	3	3	4	4	4	5	5	5
7.35	2	2	3	4	4	5	5	5	5	6
7.30	2	3	4	4	4	5	5	5	6	6
7.25	2	3	4	4	5	5	5	6	6	7
7.20	2	3	4	5	5	6	6	6	7	7
7.15	2	3	4	5	5	6	6	7	7	8
7.10	2	3	5	5	4	7	7	7	8	8
7.05	3	4	5	5	6	7	7	7	8	8
7.00	3	4	5	6	7	7	8	8	8	9

12. REFERENCES

12.1. Adams, F., and C. E. Evans. 1962. A rapid method for measuring lime requirement of red-yellow podzolic soils. Soil Sci. Soc. Am. Proc. 26: 355-357.

12.2. Hajek, B. F., F. Adams, and J. T. Cope. 1972. Rapid determination of exchangeable bases, acidity and base saturation for soil characterization. Soil Sci. Soc. Am. Proc. 36: 436-438.

DETERMINATION OF SOIL BUFFER pH BY THE SMP LIME BUFFER - ORIGINAL AND DOUBLE-BUFFER ADAPTATION

1. PRINCIPLE OF THE METHOD

1.1.   This procedure describes the determination of the lime requirement of a soil by the SMP buffer method (see 12.1, 12.2, 12.3). The lime of a requirement acid soil is defined as the amount of lime or other base which when incorporated with a given depth of acid soil increases the pH to some selected level. It is expressed as $CaCO_3$ equivalent in tons per acre of plowed soil to a depth of 8 inches (20 cm) which is equivalent to 2240 kg/ha $CaCO_3$ to the same depth or to 1.67 meq/100 g soil. An acres inches of soil is assumed to weigh 2,400,000 lbs.

1.2.   The Shoemaker, McLean, and Pratt (SMP) buffer method measures the change in pH of a buffer caused by the acids in the soil, and this change in buffer pH is a measure of the lime requirement of the soil. Th double-buffer adaptation involves the individual slope of the buffer-indicated vs. actual lime requirement curve for a given soil instead of a mean slope involved in the original method (12.1, 12.2).

2. RANGE AND SENSITIVITY

2.1.   The SMP buffer method is very reliable for soils with a greater than 4480 kg/ha (2 tons/acre) lime requirement. It is also well adapted for acid soils with a pH below 5.8 containing less than 10% organic matter, and having appreciable quantities of soluble aluminum (12.4).

2.2.   The original method was never considered to be very accurate for soils with lime requirements less than 4480 kg/ha (2 tons/acre) because of randon variation of buffer-indicated vs. actual lime requirements in this range (12.1). Also, on mineral soils of high organic matter and high levels of extractable Al the original SMP method indicates lime requirements which are lower than the actual amounts required. However, an adaptation of the original method for use on organic soils was included some time ago (see 11.3). More recently the double-buffer adaptation originally suggested by Yuan (12.5) has been developed to improve some of the shortcomings of the original method (12.2, 12.3).

3. INTERFERENCES

3.1.   Increased time of soil contact with the buffer results in lower buffer pH and therefore, a greater lime requirement.

3.2.   Organic matter and lime requirement are highly correlated (12.6), because when organic matter increases, more acidic cations accumulate on the exchange sites When organic matter is very high, especially when aluminum is complexed with it, a portion of the lime requirement may not be measured by the original SMP method (12.3).

4. SENSITIVITY

4.1.   A sensitivity of 0.1 pH unit is needed for the interpretation of this analysis by the original procedure (12.1). But the double-buffer adaptation calls for pH readings to the nearest 0.01 pH unit (12.2). A difference of 0.1 pH unit in the original method results in a difference of 0.4 to 0.6 tons of lime per acre for organic soils limed to pH 5.2 and 0.5 to 0.9 tons per acre for mineral soils limed to near neutral pH. Similarly, a difference of 0.1 pH unit in one of the two buffers in the double-buffer adaptation may result in a difference in lime requirement of less than 0.1 ton per acre for mineral soils of low lime requirement to more than 0.5 ton per acre for soils of high lime requirement.

4.2.   The SMP buffer method is not applicable to soils which have a low buffer capacity. By adding the SMP buffer solution (pH 7.5) to the soil-water suspension used to determine the pH water, the pH will increase to a level, which can not be used in combination with the table mentioned under 11.3.

5. APPARATUS

5.1.   No. 10 (2-mm opening) sieve.

5.2.   Scoop, 4.25-cm^3 volumetric.

5.3.   Cup, 50-ml (glass, plastic, or waxed paper of similar size).

5.4   Pipette, 5-ml capacity.

5.5.   Mechanical shaker.

5.6.   pH meter, line or battery operated with reproducibility to 0.01 pH unit and glass electrode paired with a calomel reference electrode.

5.7. Glassware and dispensing apparatus for the preparation and dispensing SMP buffer.

5.8. Analytical balance.

6. REAGENTS

6.1. Buffer Solutions - pH 4.0 and pH 7.0 buffers for standardization of pH meter.

6.2. SMP Buffer Solution - Weigh and transfer to an 18- liter bottle: 32.4 g paranitrophenol, 54.0 g potassium chromate ($KCrO_4$), and 955.8 g calcium chloride dihydrate ($CaCl_2 \cdot 2H_2O$). Add approximately 9 liters pure water. Shake vigorously as water is added and continue shaking for a few minutes to prevent formation of a crust over the salts. Weigh 36.0 g calcium acetate ($Ca(C_2H_3O_2) \cdot H_2O$) into a separate container and dissolve in approximately 5 liters pure water. Add latter solution to the former, shaking as they are combined. Shake every 15 or 20 minutes for 2 or 3 hours. Add 45 ml triethanolamine; again shaking as the addition is made. Shake periodically until completely dissolved. This takes approximately 8 hours. Dilute to 18 liter with distilled water. Adjust to pH 7.5 with 15% NaOH using the standardized pH meter. Filter through a fiber glass sheet or cotton mat. Connect an air inlet with 1" x 12" cylinder of drierite, 1" x 12" cylinder of ascarite, and 1"x 12" cylinder of drierite in series to protect against contamination with carbon dioxide and water vapor. (Although less tedious procedures may be used for preparing small quantities of the buffer solution, the above procedure has been found to be most satisfactory for preparing bulk quantities of the buffer solution).

7. DETERMINATION

7.1. Weigh 5 g or scoop 4.25-cm^3 of air-dry <10-mesh (2-mm) soil into a 50-ml cup in a tray designed for a mechanical shaker. Add 5 ml of pure water, shake or stir one minute, let stand 10 minutes, and read pH in water with slight swirling of the electrodes. Add 10 ml of SMP buffer adjusted to pH 7.5 (see 6.2) to the above soil suspension, shake on a mechanical shaker at 180+ oscillations per minute for 10 minutes, open the lid of the shaker, and let stand 30 minutes. Read buffer pH (pH_1) on carefully adjusted pH meter to nearest 0.01 pH unit. A 15 minute shaking time and 15 minute standing time may be used if more adaptable to the soil testing routine, since this gives essentially the same results as 10 minute shaking + 30 minute standing times.

7.11. If the original SMP (one-buffer) method is to be used, select the lime requirement from the table (see 11.3) based on the buffer pH obtained.

7.12. If the double-buffer adaptation is to be used, continue the procedure as follows: Using an automatic pipette, add to the above soil- buffer suspension an aliquot of HCl equivalent to the amount required to decrease a 10-ml aliquot of pH 7.5 buffer to pH 6.0 (1 ml of 0.206 N HCl - 0.206 meq.). Repeat the 10 minute shaking, 30 minute standing (or 15 minutes of shaking + 15 minutes standing), and reading of soil-buffer pff (pH$_2$). Use the double-buffer formula and mathematical function indicated below to convert pH readings of the quick test method to actual LR values. The procedure for use of the double-buffer adaptation has not yet been worked out for liming organic soils to pH 5.2.

7.2. COMPUTATIONS

7.21. d in meq/5 g soil =

$$\underset{(d_2)}{\Delta pH_2 \times \frac{\Delta d_2^o}{\Delta pH_2^o}} + \left[\left[\underset{(d_1)}{\Delta pH_1 \times \frac{\Delta d_1^o}{\Delta pH_1^o}} - \underset{(d_2)}{\Delta pH_2 \times \frac{\Delta d_2^o}{\Delta pH_2^o}} \right] \times \underset{(\beta)}{\left[\frac{6.5 - pH_2}{pH_1 - pH_2} \right]} \right]$$

where:

1) pH$_1$ is soil-buffer pH in pH 7.5 buffer.

2) pH$_2$ is soil-buffer pH in px 6.0 buffer.

3) $\Delta pH_1 = 7.5 - pH_1$.

4) $\Delta pH_2 = 6.0 - pH_2$.

5) $\Delta d_1^o = $ change in acidity per unit change in pH of 10 mL ΔpH^o_1 of pH 7.5 buffer by titration ~ 0.137 meq/unit pH.

6) $\Delta d^o_2 = $ change in acidity per unit change in pH of 10 mL ΔpH^o_2 of pH 6.0 buffer by titration ~ 0.129 meq/ unit pH.

7) 6:5 = pH to which soil is to be limed (any pH may be chosen).

7.22. LR in meq/100 g soil = 1.69y - 0.86 where y = 20d and d is the acidity in meq/5 g soil measured by the double-buffer (or one-buffer two-pH) procedure. The equation derived from the regression of buffer-indicated vs. actual ($Ca(OH)_2$ titrated) lime requirements corrects for less than complete reaction with the soil acidity in 10 minutes shaking and 30 minutes standing time (or 15 minutes shaking + 15 minutes standing). (If a soil is so acid that 5 g depresses the pH 6.0 buffer below pH 4.8, 4 g can be used with a multiple of 25 instead of 20.)

8. CALIBRATION AND STANDARDS

8.1. The pH meter is calibrated using prepared (see 6.1) or commercially available buffer solutions of pH 7.0 and pH 4.0 according to the instrument instruction manual.

9. CALCULATIONS

9.1. Lime requirements computed from double-buffer, quick- test formulas and expressed as meq/100 g of soil are converted to tons $CaCO_3$ per acre 8" depth of soil (2,400,000 lb) by multiplying LR by 0.6, or per acre 6 2/3" (2,000,000 lb) by multiplying by 0.5.

10. EFFECTS OF STORAGE

10.1. Air-dry soils may be stored several months in closed containers without appreciable effect on the pH_{SMP} measurement.

10.2. If the pH meter and electrodes are not to be used for extended periods of time, the instructions for storage published by the instrument manufacturer should be followed.

11. INTERPRETATION

11.1. The regular (single-buffer) SMP method is probably the most satisfactory compromise between simplicity of measurement and reasonable accuracy for soils of a wide range in lime requirement (12.3). As indicated above, the lime requirement for any soil can be determined in 11.3. This table is based on the SMP soil-buffer pH values and gives the lime requirement in terms of tons per acre of agricultural ground limestone of total neutralizing power (TNP) or $CaCO_3$ equivalent of 90% or above and an 8" plow depth (2,400,000 lb.) to increase soil pH to selected levels.

11.2. The double-buffer adaptation is somewhat more accurate for all soils, but is especially so for soils of relatively low lime requirement (12.2) and probably so for acid mineral soils of relatively high organic matter content.

11.3. Amounts of lime required to bring mineral and organic soils to indicated pH according to soil-buffer pH (tons/acre 8" soil) are presented in the following table:

Soil-buffer pH	Mineral soils				Organic soils
	7.0	7.0	6.5	6.0	5.2
	Pure CaCO3	– –	Ag-ground limestone*	– –	
6.8	1.1	1.4	1.2	1.0	0.7
6.7	1.8	2.4	2.1	1.7	1.3
6.6	2.4	3.4	2.9	2.4	1.8
6.5	3.1	4.5	3.8	3.1	2.4
6.4	4.0	5.5	4.7	3.8	2.9
6.3	4.7	6.5	5.5	4.5	3.5
6.2	5.4	7.5	6.4	5.2	4.0
6.1	6.0	8.6	7.2	5.9	4.6
6.0	6.8	9.6	8.1	6.6	5.1
5.9	7.7	10.6	9.0	7.3	5.7
5.8	8.3	11.7	9.8	8.0	6.2
5.7	9.0	12.7	10.7	8.7	6.7
5.6	9.7	13.7	11.6	9.4	7.3
5.5	10.4	14.8	12.5	10.2	7.8
5.4	11.3	15.8	13.4	10.9	8.4
5.3	11.9	16.9	14.2	11.6	8.9
5.2	12.7	17.9	15.1	12.3	9.4
5.1	13.5	19.0	16.0	13.0	10.0
5.0	14.2	20.0	16.9	13.7	10.5
4.9	15.0	21.1	17.8	14.4	11.0
4.8	15.6	22.1	18.6	15.1	11.6

*Ag-ground lime of 90% plus total neutralizingpower (TNP) or $CaCO_3$ equivalent, and fineness of 40% <100 mesh, 50% <60 mesh, 70% <20 mesh, and 95% <8 mesh.

12. REFERENCES

12.1. Shoemaker, H. E., E. O. McLean, and P. G. Pratt. 1962. Buffer methods for determination of lime requirement of soils with appreciable amount of exchangeable aluminum. Soil Sci. Soc. Am. Proc. 25:274-277.

12.2.   McLean, E. O., J. F. Trierweiler, and D. J. Eckert. 1977. Improved SMP buffer method for determining lime requirement of acid soils. Comm. Soil Sci. Plant Anal. 8:667-675.

12.3.   McLean, E. O., D. J. Eckert, G. Y Reddy, and J. F. Trierweiler. 1978. Use of double-buffer and quicktest features for improving the SMP method for determining lime requirement of acid soils. Soil Sci. Soc. Am. J. 42:311-316.

12.4.   McLean, E. O., S. W. Dumford, and F. Coronel. 1966. A comparison on several methods of determining lime requirements of soils. Soil Sci. Soc. Am. Proc. 30:26-30.

12.5.   Yuan, T. L. 1974. A double-buffer method for the determination of lime requirement of acid soils. Soil Sci. Soc. Am. Proc. 38:437-440.

12.6.   Keeney, D. R. and R. G. Corey. 1963. Factors affecting the lime requirements of Wisconsin soils. Soil Sci. Soc. Am. Proc. 27:277-280.

12.7.   Coleman, N. T. and G. W. Thomas. 1967. The basic chemistry of soil acidity. pp. 1-41. IN: R. W. Pearson and F. Adams (ed). Soil Acidity and Liming. Am. Soc. Agron., Hadison, Wis.

12.8.   Schwertmann, U. and M. L. Jackson. 1964. Influence of hydroxyaluminum ions on pk titration curves of hydronium-alumin clays. Soil Sci. Soc. Am. Proc. 28:179-182.

DETERMINATION OF EXCHANGEABLE ACIDITY AND LIME REQUIREMENT BY THE MEHLICH BUFFER-pH METHOD

1. PRINCIPLE OF THE METHOD

1.1 This procedure describes the determination of weight per volume, soil pH, and buffer pH, including the calculation of soil acidity (AC), estimation of unbuffered salt exchangeable acidity (ACe) and lime requirement (LR) (see 12.1). Methods now in use are calibrated mainly against soil pH (see 12.2), against percentage of base unsaturation (see 12.3) and use of a double buffer for sandy soils   (see 12.4, 12.5). However, in view of the importance of unbuffered salt exchangeable acidity in relation to liming (see 12.6, 12.7, 12.8), there is a need for a buffer pH method primarily calibrated against ACe and with special reference to exchangeable Al^{3+}. In addition, the buffer pH acidity (AC_b) was standardized against crop response to liming under greenhouse and field conditions (see 12.1). For the estimation of ACe, determined AC was used in regression equations for mineral soils and Histosols. The main objective of the procedure for LR was to determine the quantity of lime needed to neutralize a portion or all the ACe required for optimum plant growth. This quantity was expressed in a curvilinear function of AC for mineral soils, including those having histic epipedon and for Histosols as a function of AC used in the regression equation for calculating ACe.

1.2 Measurements of weight per volume in conjunction with the percentage of organic matter provide the indexes for differentiating between mineral soils and Histosols for calculating LR. The determination of soil pH provides the index to the need for liming depending on acid (Al^{3+}) tolerance of crops and major soil differences.

2. RANGE AND SENSITIVITY

2.1 The capacity of the buffer (pH 6.6-4.0) is equivalent to 10 meq $CaCO_3$/100 cm^3, 20 metric tons (MT) of limestone/ha or 20,000 lb/A. Provisions in the procedure allow this capacity to be doubled.

2.2 Sensitivity per 0.1 pH depression of buffer is 0.4 meq $CaCO_3$/100 cm^3, 0.4 MT/ha or 400 lb/A.

3. INTERFERENCES

3.1   Buffer pH of soil suspension should be read after 60 minutes standing. All measurements should be made within the same day.

4. SENSITIVITY

4.1   A sensitivity of 0.1 pH unit is adequate for soils having a LR greater than 2 MT/ha, while for a lower LR, a sensitivity of 0.05 pH units would be desirable.

5. APPARATUS

5.1   No. 10, 2-mm ISO standard sieve.

5.2   10-cm^3 scoop, volumetric.

5.3   50-mL cup, glass, plastic or waxed paper cup.

5.4   Pipette, 10-mL capacity.

5.5   Mechanical shaker or stirrer (optional).

5.6   pH meter, line or battery operated, with reproducibility to 0.05 pH unit, and a glass electrode paired with a calomel reference electrode.

5.7   Glassware and dispensing apparatus for the preparation and dispensing of 10 mL H_2O and Mehlich buffer reagent.

5.8   Analytical balance.

6. REAGENTS

6.1   <u>Sodium glycerophosphate (National Formulary) (N.F.) M.W. 315.15)</u> - The N. F. quality of sodium glycerophosphate [$Na_2C_3 H_5 (OH)_2 PO_4 \cdot 5\ 1/2\ H_2O$] is very satisfactory and considerably more economical than the crystal Beta form. (Source: Roussel Corporation, 155 E. 44th St., New York, NY 10017).

6.2   <u>Calibration Buffer Solutions</u> - pH 4.0 and pH 7.0 for standardization of the pH meter-electrode system (see 6.1, 6.2, page 2).

6.3   Buffer Solution - To about 1500 mL pure water in a 2-liter flask or a 2-liter calibrated bottle, add 5 mL glacial acetic acid ($H_2 C_2 H_3 O_2$) and 9 mL triethanolamine or, for ease of delivery, add 18 mL of an 1:1 aqueous mixture . Add 86 g ammonium chloride (NH_4Cl) and 40 g barium chloride ($BaCl_2 \cdot 2H_2O$) and dissolve. Dissolve separately 36 g sodium glycero-phosphate in 400 mL pure water and transfer to the above 2-liter flask or bottle. Allow the endothermic reacted solution to reach room temperature and make up to volume with pure water and mix. Dilute an aliquot of the Buffer Solution with an equal volume of pure water and determine the pH. The pH of the Buffer Reagent should be 6.6. However, if it is above pH 6.64, add dropwise glacial acetic acid. If it is below pH 6.56, add dropwise 1:1 aqueous triethanolamine. Check the concentration of the buffer by adding 10 mL 0.1N HCl-AlCl$_3$ mixture [dissolve 4.024 g aluminum chloride ($AlCl_3 \cdot 6H_2O$) in 0.05N hydrochloric acid (HCl)] to 10 mL buffer + 10 mL pure water and determine the pH. The correct pH obtained should be 4.1 $\pm$ 0.05.

7. PROCEDURE

7.1   Scoop 10-cm^3 of air-dry soil, <10mesh (2-mm), into a 50-mL cup (see 5.2, 5.3). To obtain weight per volume, weigh the measured 10-cm^3 soil to the nearest 0.1 g, divide by 10 and express the results in g/cm^3.

7.2   Add 10 mL pure water with sufficient force to mix with soil. After stirring for about 30 minutes, read soil pH while stirring (for poorly wettable Histosols, add 8 to 10 drops of ethanol).

7.3   Add to the soil from the pH determination 10 mL Buffer Solution (see 6.3) with sufficient force to mix. Read the buffer pH to the nearest 0.05 unit after 60 minutes while stirring. If it is desired to extend buffer capacity below pH 4.0, add an additional 10 mL Buffer Solution (see 6.3), equilibrate with stirring and measure pH$_B$.

8. CALIBRATION AND STANDARDS

8.1   Before measuring the buffer pH of the soil suspension, calibrate the pH meter to pH 6.6 in a mixture of 10 mL Buffer Solution (see 6.3) and 10 mL pure water.

9. CALCULATIONS FOR EXCHANGEABLE ACIDITY

9.1   Convert buffer pH (BpH) into buffer pH acidity (AC) as follows:

$$AC \text{ in meq/100 cm}^3 \text{ soil} = (6.6 - BpH)/0.25$$

(If a second 10-mL portion of buffer was used, multiply AC by 2.)

9.2   For unbuffered salt exchangeable acidity (ACe) based on AC of mineral soils, calculate:

$$ACe \text{ in meq/100 cm}^3 = 0.54 + 0.96 \text{ (AC)}$$

9.3   For ACe determination of Histosols and mineral soils having histic epipedon, calculate:

$$ACe \text{ in meq/100 cm}^3 = -7.4 + 1.6 \text{ (AC)}$$

10. CALCULATIONS FOR LIME REQUIREMENT

10.1   Convert BpH into AC (see 9.1).

10.2   The Lime Requirement (LR) in the following equations may be expressed, and is equivalent to meq $CaCO_3$/100 cm^3 soil, metric tons (MT) ground limestone TNP = 90%/ha to a depth of 20 cm, or lbs/A (MT x 10^3).

10.3   Mineral Soils - For plants with slight to moderate tolerance for ACe and soil reaction in H_2O pH 5.8 to 6.5,

$$LR = 0.1 \text{ (AC)}^2 + AC$$

10.4   Mineral Soils - For plants with low tolerance to ACe, and soil reaction in H_2O < 6.5, multiply results with equation [3] by 1.5 or 2.0.

10.5   Histosols or Mineral Soils with Histic Epipedon (OM 20% and above) - For soil reaction in H_2O < pH 4.8 to 5.0 and W/V in g/cm^3 0.75 , use equation 2 x 1.3, viz.,

$$LR = [-7.4 + 1.6(AC)] \text{ 1.3}$$

10.6   Mineral Soils High in Organic Matter (OM 10%-19%) - For soil reaction in H_2O < pH 5.3 to 5.5, and W/V within 0.75 to 0.95 g/cm^3, use equation [4] with soils of sandy texture and equation [3] with soils of silt and clay texture.

10.7 In all cases, when soil pH is below the indicated optimum, we suggest the use of 1 ton limestone/ha or its equivalent, even though AC is less than 0.5 meq/100 cm³.

11. INTERPRETATION

11.1 The LR equations based on the proposed BpH method may be used in a computerized soil testing program. For manual use, the calculated LR values based on equations [3] and [4] at 0.1 BpH intervals are recorded in the following table.

Buffer pH, AC and LR Conversion of Mineral and Organic Soils into MT/ha or lbs/A (MT x 10³) of Ag. Ground Limestone with TNP = 90%

		LR for soils				LR for soils	
		Mineral	Organic			Mineral	Organic
		equation	equation			equation	equation
BpH	AC	[3]*	[4]*	BpH	AC	[3]*	[4]*
6.6	0.0	0.0	0.0	5.2	5.6	8.7	2.0
6.5	0.4	0.4**	0.0	5.1	6.0	9.6	2.9
6.4	0.8	0.9	0.0	5.0	6.4	10.5	3.7
6.3	1.2	1.3	0.0	4.9	6.8	11.4	4.5
6.2	1.6	1.9	0.0	4.8	7.2	12.4	5.4
6.1	2.0	2.4	0.0	4.7	7.6	13.4	6.2
6.0	2.4	3.0	0.0	4.6	8.0	14.4	7.0
5.9	2.8	3.6	0.0	4.5	8.4	15.5	7.9
5.8	3.2	4.2	0.0	4.4	8.8	16.5	8.7
5.7	3.6	4.9	0.0	4.3	9.2	17.7	9.5
5.6	4.0	5.6	0.0	4.2	9.6	18.8	10.3
5.5	4.4	6.3	0.0	4.1	10.0	20.0	11.2
5.4	4.8	7.1	0.4**	4.0	10.4	21.2	12.0
5.3	5.2	7.9	1.2	3.9	10.8	22.5	12.8

* For equations [3] and [4], see 10.3 to 10.6. For crops with high LR or very low tolerance to ACe, multiply the results of equation [3] by a factor of 1.5 or 2.0.

** We suggest using 1 ton limestone/ha or 1000 lbs/A when LR based on pH is indicated (see 10.3 to 10.6).

11.2 While liming needs are contingent on BpH, soil pH measured in a 1:1 soil:water ratio on a volume basis has been suggested as a criterion in the LR decision-making process. Soil pH levels measured in N KCl and 0.01M CaCl₂ were found to deviate inconsistently from those measured in H₂O.

These deviations were largely related to the quantity and proportion of ACe to ACr, exchangeable Al $^{3+}$ to H^+ and major soil components with respect to organic matter, layer silicates and sesquioxides hydrates. Schofield and Taylor (see 12.9) introduced the use of 0.01M $CaCl_2$ in a 1:2 soil:salt solution ratio on a weight-to-volume basis as a measure of "lime potential." The authors determined pH in the supernatant liquid. Jackson (see 12.10) stirred the soil suspension just before immersing the electrodes, and Peech (see 12.11) placed the glass electrode into the partly settled suspension and the calomel electrode into the clear supernatant solution. In the case of acid Ultisols, the relative decrease in pH from that obtained in a 1:1 soil: water suspension was, on the average, 1.0, 0.8 and 0.6 by the Jackson, Peech and Schofield-Taylor procedures, respectively. With neutral to slightly acid soils, the total differences were in general less than one-half of the above. In view of the variability of soil pH obtained with varying salts due to procedural differences and soil properties, and because of the importance of maintaining uniformity of soil test results, measurement of pH in a 1:1 soil:water suspension by volume in conjunction with the proposed BpH method for LR is recommended.

12. REFERENCES

12.1. Mehlich, A. 1976. New buffer pH method for rapid estimation of exchangeable acidity and lime requirement of soils. Comm. Soil Sci. Plant Anal. 7: 637-652.

12.2 Shoemaker, H. E., E. O. McLean, and P. F. Pratt. 1962. Buffer methods for determination of lime requirement of soils with appreciable amount of exchangeable aluminum. Soil Sc. Soc. Am. Proc. 25:274-277.

12.3. Adams, F., and C. E. Evans. 1962. A rapid method for measuring lime requirement of red-yellow podzolic soils. Soil Sc. Soc. Am. Proc. 26: 355-357.

12.4. Yuan, T. L. 1974. A double buffer method for the determination of lime requirement of acid soils. Soil Sc. Soc. Am. Proc. 38:437-440.

12.5 Yuan, T. L. 1975. Lime requirement determination of sandy soils by different rapid methods. Soil and Crop Sci. Soc. Fla. Proc. 25:274-277.

12.6 Mehlich, A., S. S. Bowling, and A. L. Hatfield. 1976. Buffer pH acidity in relation to nature of soil acidity and expression of lime requirement. Comm. Soil Sci. Plant Anal. 7:253-263.

12.7   Kamprath, E. J. 1970. Exchangeable aluminum as a criterion
        for liming leached mineral soils.   Soil Sc. Soc. Am. Proc.
        34:252-254.

12.8   Evans, C. E., and E. J. Kamprath. 1970. Lime response as
        related to percent Al saturation, solution Al and organic matter
        content. Soil Sci. Soc. Am. Proc. 34:893-896.

12.9   Schofield, R. K., and A. W. Taylor. 1955. The measurement
        of soil pH. Soil Sci. Soc. Am. Proc. 19: 164-167.

12.10   Jackson, M. 1958. Soil Chemical Analysis. Prentice-Hall,
        Inc., Englewood Cliffs, NJ.

12.11   Peech, M. 1965. Hydrogen-ion activity, pp. 914-926. IN: C. A.
        Black (ed.), Methods of Soil Analysis, Part 2, Agronomy 9.
        American Society of Agronomy, Madison, WI.

DETERMINATION OF PHOSPHORUS BY BRAY P1 EXTRACTION

1. PRINCIPLE OF THE METHOD

1.1 The extraction of phosphorus by the Bray P1 method is based upon the solubilization effect of the H+ on soil phosphorus and the ability of the F- to lower the activity of Al+3 and to a lesser extent that of Ca+2 and Fe+3 in the extraction system. As described in this section, clay soils with a moderately high degree of base saturation or silty clay loam soils that are calcareous or have a very high degree of base saturation will lessen the solubilizing ability of the extractant. Consequently, the method should normally be limited to soils with pH_w values less than 6.8 when the texture is silty clay loam or finer. Calcareous soils, or high pH, fine textured soils may be tested by this method, but higher ratios of extractant-to-soil are often used for such soils (see 12.6). Another alternative is the Olsen P procedure, page 47. The Bray P1 method is also suitable for organic soils.

1.2 The extractant was developed and first described by Bray and Kurtz (see 12.1). The extraction time and the solution-to-soil ratio in their procedure were 1 minute and 7 mL extractant to 1.0 g soil, respectively. To simplify adaptation to routine laboratory work and to extend the range of soils for which the extractant is suitable, both the extraction time and the solution to soil ratio have been altered to 5 minutes and a 1:10 soil:extractant ratio. This modification is in wide use in laboratories of the mid-east, mid-south and north central areas of the United States (see 12.8).

2. RANGE AND SENSITIVITY

2.1 This procedure yields a standard curve that is essentially linear to 10 mg/L of phosphorus in the soil extract (approximately 200 kg/ha or 178 lb/A of extractable phosphorus).

2.2 The sensitivity is approximately 0.15 mg/L in the extract (2.7 lb/A or 3.0 kg/ha phosphorus in the soil).

3. INTERFERENCES

3.1 Arsenic - Concentrations of up to 1 mg/L arsenic in the extract do not interfere (see 12.3). Jackson describes techniques for removal (see 12.5).

3.2   Silica - Silica will not interfere at <10 mg/L in the extract (see 12.3).

3.3   Fluoride - The fluoride in the extract normally will not interfere in the formation of the molybdenum blue color with phosphorus. Any interference may be eliminated with the addition of boric acid (see 12.2). Maximum color development is slower in the presence of the fluoride ion.

4. PRECISION AND ACCURACY

4.1   The reproducibility of determinations by this procedure depends upon the extent to which the times of extraction, filtration and color development are controlled. Reasonable control and thorough sample preparation should give a coefficient of variation of about 5%.

5. APPARATUS

5.1   No. 10 (2-mm opening) sieve.

5.2   Scoop, 1.70-cm^3 volumetric.

5.3   Extraction bottle or flask, 50 mL with stoppers.

5.4   Mechanical reciprocating shaker, minimum of 180 oscillations per minute.

5.5   Filter funnel, 11 cm.

5.6   Whatman No. 2 filter paper or equivalent, 11 cm.

5.7   Spectrophotometer suitable for measurement in the 880-nm range.

5.8   Spectrophotometer tube or cuvet.

5.9   Funnel racks.

5.10   Volumetric flasks and pipettes as required for preparation of reagents, standard solutions and color development.

5.11   Analytical balance.

6. REAGENTS

6.1   Extracting Reagent (0.03N NH$_4$ F in 0.025N HCl) :

6.11   1N NH$_4$F - Dissolve 37 g ammonium fluoride in 400 mL pure water and dilute the solution to 1 liter. Store in a polyethylene container and avoid prolonged contact with glass.

6.12   0.5N HCl - Dilute 20.4 mL conc HCl to 500 mL with pure water.

6.13   Extracting Reagent - Mix 30 mL 1N NH$_4$F (see 6.11) with 50 mL of 0.5N HCl (see 6.12) and dilute to 1 liter with pure water. This solution is 0.03N in NH$_4$ F and 0.02N in HCl and has a pH of 2.6. Stored in polyethylene, it is stable for more than 1 year.

6.2   Ascorbic Acid Solution - Dissolve 132.0 g ascorbic acid in pure water and dilute to 1 liter with pure water. Store in dark glass bottle in a refrigerated compartment.

6.3   Sulfuric-Molybdate Solution - Dissolve 60 g ammonium molybdate [(NH$_4$)$_6$Mo$_7$O$_{24}$ · 4H$_2$O] in 500 mL of pure water. Dissolve 1.455 g antimony potassium tartrate [K(SbO)C$_4$H$_4$O$_6$.1/2H$_2$O] in the molybdate solution. Add slowly 700 mL conc H$_2$SO$_4$ and mix well. Let it cool and dilute to 1 liter with pure water. This solution may be blue but will produce a clear solution when the Working Solution (see 6.4) is prepared. Store in a polyethylene or pyrex bottle in a dark, refrigerated compartment.

6.4   Working Solution - Add 10 mL ascorbic acid solution (see 6.2) to about 800 mL pure water followed by 25 mL Sulfuric-Molybdate Solution (see 6.3), and then dilute to 1 liter with pure water. Allow to stand at least 1 hour before using. Prepare fresh daily.

6.5   Phosphorus Standard (100 mg/L) - Weigh 0.4394 g monobasic potassium phosphate (KH$_2$PO$_4$) which has been oven-dried at 100° C into a 1-liter volumetric flask and bring to volume with Extracting Reagent. (see 6.13).

7. PROCEDURE

7.1   Extraction - Weigh 2.0 g or scoop 1.70-cm^3 air-dry <10-mesh (2-mm) soil into a 50-mL extraction bottle or flask, add 20 mL Extracting Reagent (see 6.13) and shake for 5 minutes on a reciprocating shaker (see 5.4). Filter through Whatman No. 2 filter paper (see 5.6), limiting the filtration time to 10 minutes, and save the extract.

7.2 <u>Color Development</u> - Transfer exactly 2.0 mL extract or standard solution to a spectrophotometer tube or a cuvet (see 5.8). Add 8 mL of Working Solution (see 6.4) and mix the contents of the tube thoroughly. After 10 minutes, measure the percentage of transmittance (%T) at 882 nm. The color intensity is stable for 4 hours.

8. CALIBRATION AND STANDARDS

8.1 <u>Working Phosphorus Standards</u> - With the Standard Phosphorus Solution (see 6.5), prepare 6 Working Standard Solutions containing from 0.2 to 10 mg/L of phosphorus in the final volume. Make all dilutions with Extracting Reagent (see 6.13).

8.2 <u>Calibration Curve</u> - On semilog graph paper, plot the percentage of transmittance (%T) on the logarithmic scale versus mg/L phosphorus in the standard solutions on the linear scale.

9. CALCULATION

9.1 The results are reported as kg P/ha for a 20-cm depth of soil. Kg/ha of phosphorus in the soil = mg/L in the extractant x 22.4, lbs P/A = mg/L in the extractant x 20.

(Note: This assumes that a uniform 2.0 mL aliquot is used for standards and unknowns in 7.2.)

10. EFFECTS OF STORAGE

10.1 After air drying, the extractable phosphorus levels in soils remain stable for several months.

10.2 After extraction, the phosphorus in the extract should be measured within 72 hours.

11. INTERPRETATION

11.1 Accurate fertilizer recommendations for phosphorus must be based on field response data conducted under local soil-climate-crop conditions (see 12.7). In general, the extractable phosphorus levels may be categorized as follows:

	Extractable P	
<u>Category</u>	<u>kg/ha</u>	<u>lb/A</u>
Low	<34	(<30)
Medium	34-68	(30 - 60)
High	>68	(>60)

12. REFERENCES

12.1. Bray, R. H., and L. T. Kurtz. 1945. Determination of total, organic and available forms of phosphorus in soils. Soil Sci. 59:39-45.

12.2. Kurtz, L. T. 1942. Elimination of fluoride interference in molybdenum blue reaction. Ind. Eng. Chem. Anal. Ed. 14:855.

12.3. Murphy, J., and J. P. Riley. 1962. A modified single solution method for the determination of phosphate in natural waters. Anal. Chem. Acta 27:31-36.

12.4. Olsen, S. R., and L. A. Dean. 1965. Phosphorus, pp. 1035-1949. IN: C.A. Black (ed.), Methods of Soil Analysis, Part 2, Agronomy No. 9. American Society of Agronomy, Madison, WI.

12.5 Jackson, M. L. 1958. Soil Chemical Analysis. Prentice-Hall, Inc., Englewood Cliffs, NJ.

12.6 Smith, F. W., B. G. Ellis, and J. Grava. 1957. Use of acid-fluoride solutions for extraction of available phosphorus in calcareous soils and in soils which rock phosphate has been added. Soil Sci. Soc. Am. Proc. 21:400-404.

12.7 Thomas, G. W., and D. E. Peaslee. 1973. Testing soil for phosphorus, pp. 115-132. IN: L. M. Walsh and J. D. Beaton (eds.), Soil Testing and Plant Analysis, rev. ed. Soil Science Society of America, Madison WI.

12.8 Knudsen, D. 1980. Recommended phosphorus tests, pp. 14-16. IN: W. C. Dahnke (ed.), Recommended Chemical Soil Test Procedure for the North Central Region. North Central Regional Publication No. 221 (rev.). North Dakota Agricultural Experiment Station, Fargo, ND.

DETERMINATION OF PHOSPHORUS BY OLSEN'S SODIUM BICARBONATE EXTRACTION

1. PRINCIPLE OF THE METHOD

1.1 The extractant is a 0.5M Na bicarbonate ($NaHCO_3$) solution at a pH of 8.5 (see 12.1). The solubility of calcium phosphate in calcareous, alkaline or neutral soils is increased because of the precipitation of Ca^{++} as $CaCO_3$. In acid soils, phosphorus concentration in solution increases when aluminum and iron phosphates such as variscite and strengite are present (see 12.2). Secondary precipitation reactions are reduced in acid and calcareous soils because iron, aluminum and calcium concentrations remain low in the extract (see 12.3).

1.2 The extractant was first developed and described by Olsen et al. (see 12.1). The original procedure required that 5 g soil be shaken for 30 minutes in 100 mL extraction reagent containing 1 teaspoon of carbon black (Darco G-60). The use of carbon black eliminated the color in the extract. This procedure was recently modified so that the use of carbon black was eliminated (see 12.4). In the modified method, a single solution reagent which consists of an acidified solution of ammonium molybdate containing ascorbic acid and a small amount of antimony is used (see 12.4 and 12.5).

2. RANGE AND SENSITIVITY

2.1 The molybdate complex obeys Beer's Law, yielding a straight line when the log of % transmittance vs. concentration is plotted--up to a concentration of 2 mg P/L in the final solution (see 12.5).

2.2 The sensitivity of this method is 0.02 mg P/L in the extract (see 12.5).

3. INTERFERENCES

3.1 This method provides non-interference of silicate in solution up to at least 10 mg Si/L, up to at least 50 mg Fe (III)/L, up to at least 10 mg Cu (II)/L, and up to at least 1 mg arsenate/L (see 12.5).

4. PRECISION

4.1 The coefficient of variation (CV) depends on the concentration of phosphorus in the extract. For routine analysis, the CV may vary from 5% to 10%.

5. APPARATUS

5.1   No. 10 (2-mm opening) sieve.

5.2   2-cm^3 scoop, volumetric.

5.3   250-mL extraction bottle or flask, with stoppers.

5.4   Mechanical reciprocating shaker, 180 oscillations per minute.

5.5   Filter funnel, 11 cm.

5.6   Whatman No. 40 filter paper (or equivalent), 11 cm.

5.7   Spectrophotometer suitable for measurement at 880 nm range.

5.8   Spectrophotometric tube or cuvet.

5.9   Funnel racks.

5.10   Analytical balance.

5.11   Volumetric flasks and pipettes as required for preparation of reagents, standard solutions and color development.

6. REAGENTS

6.1   <u>Extraction Reagent</u> - Dissolve 42.0 g sodium bicarbonate ($NaHCO_3$) in pure water and dilute to 1 liter. Adjust the pH to 8.5 with 50% NaOH. Add mineral oil to avoid exposure of the solution to air. Store in a polyethylene container and check the pH of the solution each month.

6.2   <u>Mixed Reagent</u> - Dissolve 12.0 g ammonium molybdate [$(NH_4)_6 Mo_7 O_{24} \cdot 4H_2O$] in 250 mL pure water. Dissolve 0.2908 g antimony potassium tartrate [$K (SbO) C_4H_4O_6 \cdot 1/2H_2O$] in 1000 mL of 5N sulfuric acid (H_2SO_4) (148 mL conc H_2SO_4 per liter). Mix the two solutions together thoroughly and make to a 2000-mL volume with pure water. Store in a pyrex bottle in a dark, cool place.

6.3   <u>Color-Developing Reagent</u> - Add 0.739 g ascorbic acid to 140 mL Mixed Reagent (see 6.2). This amount of reagent is enough for 24 phosphorus determinations, allowing 20 mL for wastage. This reagent should be prepared as required, as it will not keep for more than 24 hours.

6.4   Phosphorus Standard (100 mg P/L) - Weigh 0.4394 g mono-basic potassium phosphate (KH_2PO_4) which has been oven-dried at $110°$ C into a 1-liter volumetric flask. Add about 200 mL pure water to dissolve the salt, and bring to volume with pure water. Add 5 drops of toluene and shake the flask vigorously. This solution contains 0.1 mg P/mL.

7. PROCEDURE

7.1   Extraction - Weigh 2.5 g or scoop 2-cm^3 (see 5.2) air-dry, <10-mesh (2-mm) soil into a 250-mL extraction bottle (see 5.3). Add 50 mL Extraction Reagent (see 6.1) and shake for 30 minutes on a reciprocating shaker (see 5.4). Filter and collect the filtrate.

7.2   Color Development - Pipette 5 mL of extract or standard solution into spectrophotometric tubes which are optically similar (see 5.8). Add 5 mL Color-Developing Reagent (see 6.3) carefully to prevent loss of the sample due to excessive foaming. Add 15 mL pure water with a burette and stir. Let it stand 15 minutes and measure the color intensity (%T) at 880 nm. If the color is too intense for the working range, reduce the aliquot from 5 mL to a lower volume and add enough Extraction Reagent (see 6.1) to bring the aliquot volume to 5 mL (if 1 mL aliquot is used, 4 mL Extraction Reagent is added to the aliquot). Develop the color. The color is stable for 24 hours.

8. CALIBRATION AND STANDARDS

8.1   Working Phosphorus Standards

8.11   Dilute 20 mL phosphorus standard (see 6.4) to 1 liter with pure water to obtain a solution containing 2 mg P/L.

8.12   Pipette aliquots of dilute standard (see 8.11) containing from 2 to 25 µg P (1 mL to 12.5 mL) into 25 mL volumetric flasks. Add 5 mL Color-Developing Reagent (see 6.3) and mix. Bring to volume with the Extraction Reagent and mix thoroughly. Let stand 15 minutes and read the transmittance at 880 nm.

8.2   Calibration Curve - On semi-log graph paper, plot the percentage of transmittance (% T) on the logarithmic scale versus mg P/L in the solution on the linear scale. Construct a table from the calibration curve showing the relationship between the % T and P concentration. Check 2 points on the calibration curve on every day of phosphorus analysis.

9. CALCULATION

9.1   The results are reported as kg P/ha for a 20-cm depth of soil. Kg/ha of phosphorus in the soil = mg/L in the final solution x 250. These calculations are for a 5-mL aliquot of extract and should be adjusted for other aliquots. To express the results on a soil-weight basis, use the following formula:

$$\text{mg P/kg in soil} = \text{mg P/L in solution} \times 100.$$

10. EFFECTS OF STORAGE

10.1   Air-drying and storage may have some effect on $NaHCO_3$ - extractable phosphorus. However, for routine soil testing purposes, this effect is not significant.

10.2   After extraction, measure the phosphorus in the extract within 2 hours.

11. INTERPRETATION

11.1   It has been shown by several workers (see 12.3) that a phosphorus content of <12 kg/ha in soil indicates a response to phosphorus fertilizers, between 12 and 24 kg/ha indicates a probable response, and >24 kg/ha indicates a response is unlikely. However, differences in climatic conditions and crop species may make the general guidelines given above not applicable to all conditions.

12. REFERENCES

12.1.   Olsen, S. R., C. V. Cole, F. S. Watanabe, and L. A. Dean. 1954. Estimation of available phosphorus in soils by extraction with sodium bicarbonate. USDA Cir. No. 939.

12.2.   Lindsay, W. L., and E. C. Moreno. 1960. Phosphate phase equilibria in soils. Soil Sci. Soc. Am. Proc. 24:177-182.

12.3.   Olsen, S. R., and L. A. Dean. 1965. Phosphorus, pp. 1044-1046. IN: C. A. Black (ed.), Methods of Soil Analysis, Agronomy No. 9. American Society of Agronomy, Madison, WI.

12.4.   Watanabe, F. S., and S. R. Olsen. 1965. Test of an ascorbic acid method for determining phosphorus in water and $NaHCO_3$ extracts from soil. Soil Sci. Soc. Am. Proc. 29:677-678.

12.5   Murphy, J., and J. P. Riley. 1962. A modified single solution method for the determination of phosphate in natural waters. Anal. Chem. Acta 27:31-36.

DETERMINATION OF POTASSIUM, MAGNESIUM, CALCIUM AND SODIUM BY NEUTRAL NORMAL AMMONIUM ACETATE EXTRACTION

1. PRINCIPLE OF THE METHOD

1.1   This method uses a neutral salt solution to replace the cations present on the soil exchange complex; therefore, the cation concentrations determined by this method are referred to as "exchangeable" for non-calcareous soils. For calcareous soils, the cations are referred to as "exchangeable plus soluble."

1.2   The use of neutral normal ammonium acetate to determine exchangeable potassium was first described by Prianischnikov (see 12.1). Schollenberger and Simon (see 12.2) describe the advantages of this extracting reagent as to its effectiveness in wetting soil, replacing exchangeable cations, ease of volatility during analysis, and suitability for use with flame emission. More recently, this method was described by Jackson (see 12.3), Chapman (see 12.4) and Hesse (see 12.5). The neutral normal ammonium acetate extraction procedure is the most commonly used extraction reagent for determining potassium, magnesium, calcium and sodium in soil-testing laboratories in the United States (see 12.6).

2. RANGE AND SENSITIVITY

2.1   The range of detection will depend on the particular instrument setup. The range can be extended by diluting the extract.

2.2   The sensitivity will vary with the type of instrument used, wavelength selected and method of excitation.

2.3   The commonly used methods of analysis are flame emission and atomic absorption spectrometry. For a more complete description of these methods, refer to Isaac and Kerber (12.7).

3. INTERFERENCES

3.1   Under certain conditions, the Extracting Reagent (see 6.2) will extract more potassium, magnesium, calcium and/or sodium than is found in exchangeable form, such as those elements released by weathering action during the period of extraction, and calcium and magnesium released through the dissolution of the carbonate form of these elements. The additional amounts of these elements will not normally significantly alter an analysis of soil fertility. If the CEC is estimated by cation summation and the percentage of base saturation is used to assess fertility status, then the interference is significant.

3.2   Known interferences and compensation for the changing characteristics of the extract to be analyzed must be acknowledged. The use of internal standards such as lithium and compensating elements such as lanthanum are essential in most flame methods of excitation (see 12.7).

4. PRECISION AND ACCURACY

4.1   Repeated analysis of the same soil with medium concentration ranges of potassium, calcium, magnesium and sodium will produce coefficients of variations of 5% to 10%. A major portion of the variance is related to the heterogeneity of the soil rather than to the extraction or method of analysis.

4.2   The level of exchangeable potassium will increase upon the air drying of some soils (see 12.8). Soil samples can be extracted in the moist state. However, the difficulties inherent in handling and storing moist soil hinder the easy adaptation of this method to a routine method of analysis. Compensation can be made, based on the expected release of potassium by the particular soil being tested.

5. APPARATUS

5.1   No. 10 (20-mm opening) sieve.

5.2 4.25-cm^3 scoop, volumetric.

5.3   50-mL extraction bottle or flask, with stoppers.

5.4   Mechanical reciprocating shaker, 180 oscillations per minute.

5.5   Filter funnel, 11 cm.

5.6   Whatman No. 1 filter paper (or equivalent), 11 cm.

5.7   Flame emission and/or atomic absorption spectrophotometer.

5.8   Funnel racks.

5.9   Analytical balance.

5.10   Volumetric flasks and pipettes as required for preparation of reagents and standard solutions.

6. REAGENTS

6.1   Extraction Reagent - Dilute 57 mL of glacial acetic acid (NC_2 H_3O_2) with pure water to a volume of approximately 500 mL. Then add 69 mL conc ammonium hydroxide (NH_4OH). CAUTION: Use fumehood. Add sufficient pure water to obtain a volume of 990 mL. Adjust the pH, after thoroughly mixing the solution to pH 7.0, using either ammonium hydroxide or acetic acid. Dilute to a final volume of 1000 mL with pure water. Alternate method: Weigh 77.1 g ammonium acetate ($NH_4C_2H_3$ O_2) in about 900 mL pure water. Adjust the pH to 7.0, after thoroughly mixing the solution, using either 3N acetic acid or 3N ammonium hydroxide. Dilute to a final volume of 1000 mL with pure water.

6.2   Potassium Standard (1000 mg/L) - Weigh 1.9080 g potassium chloride (KCl) into a 1-liter volumetric flask and bring to volume with Extraction Reagent (see 6.1). Prepare the working standards by diluting aliquots of the stock solution standard with Extraction Reagent (see 6.1) to cover the anticipated range in concentration to be found in the soil extraction filtrate. Working standards from 5 to 100 mg K/L should be sufficient for most soils.

6.3   Calcium Standard (1000 mg/L) - Weigh 2.498 g calcium carbonate ($CaCO_3$) into a 1-liter volumetric flask. Add 50 mL of pure water and add dropwise a minimum volume (approximately 20 mL) of conc hydrochloric acid (HCl) to effect the complete solution of the calcium carbonate. Dilute to the mark with Extraction Reagent (see 6.1). Prepare the working standards by diluting aliquots of the stock solution standard with Extraction Reagent (see 6.1) to cover the anticipated range in concentration to be found in the soil extraction filtrate and to fit the working range of the instrument.

6.4   Magnesium Standard (1000 mg/L) - Weigh 1.000 g magnesium ribbon into a 1-liter volumetric flask and dissolve in the minimum volume of dilute hydrochloric acid (HCl). Dilute to 1 liter with Extraction Reagent (see 6.1). Prepare working standards by diluting aliquots of the stock solution standard with Extraction Reagent (see 6.1) to cover the anticipated range in concentration to be found in the soil extraction filtrate and to fit the optimum range of the instrument.

6.5   Sodium Standard (1000 mg/L) - Weigh 2.542 g sodium chloride (NaCl) into a 1-liter volumetric flask and bring to volume with Extraction Reagent (see 6.1). Prepare working standards by diluting aliquots of the stock solution standards with Extraction Reagent (see 6.1) to cover the anticipated range in concentration to be found in the soil extraction filtrate. Working standards from 1 to 10 mg Na/L should be sufficient for most soils.

7. PROCEDURE

7.1   Extraction - Weigh 5 g or scoop 4.25-cm^3 (see 5.2) of air-dry <10-mesh (2-mm) soil into a 50-mL extraction bottle (see 5.3). Add 25 mL of Extraction Reagent (see 6.1) and shake for 5 minutes on a reciprocating shaker (see 5.4). Filter and collect the filtrate.

7.2   Analysis - The elements potassium, magnesium, calcium and sodium in the filtrate can be determined by either flame emission or atomic absorption spectrometry. Since instruments vary in their operating conditions, no specific details are given here. It is recommended that the procedures described by Isaac and Kerber (see 12.7) be followed.

8. CALIBRATION AND STANDARDS

8.1   Working Standards - Working standards should be prepared as described in section 6. If element concentrations are found outside the range of the instrument or standards, suitable dilutions should be prepared, starting with a 1:1 soil extract-to-Extracting Reagent (see 6.1) dilution. Dilution should be made so as to minimize the magnification of errors introduced by diluting.

8.2   Calibration - Calibration procedures vary with instrument techniques and the type of instrument. Every precaution should be taken to use proper procedures and to follow the manufacturer's recommendations in the operation and calibration of the instrument.

9. CALCULATION

9.1   The results are reported as kg/ha for a 20-cm depth of soil. Kg element/ha = mg/L of element in extraction filtrate x 10.

9.2   To convert to other units for comparison, see Mehlich (12.9).

10. EFFECTS OF STORAGE

10.1   Soils may be stored in an air-dry condition for several months with no effects on the exchangeable potassium, magnesium, calcium and sodium content. Potassium may be released on drying for some soils (see 12.8).

10.2   After extraction, the filtrate containing potassium, magnesium, calcium and sodium should not be stored for longer than 24 hours unless it is refrigerated or treated to prevent bacterial growth.

11. INTERPRETATION

11.1   An evaluation of the analytical results for determination of fertilizer recommendations, particularly for the elements potassium and magnesium, must be based on field response data conducted under local soil-climate-crop conditions (see 12.9 and 12.10).

12. REFERENCES

12.1   Prianischnikov, D. N. 1913. Landw. vers. Sta. 79-80:667.

12.2   Schollenberger, C. J., and R.H. Simon. 1945. Determination of exchange capacity and exchangeable bases in soil-ammonium acetate method. Soil Sci. 59:13-24.

12.3   Jackson, M. L. 1958. Soil Chemical Analysis. Prentice-Hall, Inc., Englewood Cliffs, NJ.

12.4   Chapman, D. D. 1965. Total exchangeable bases, pp. 902-904. IN: C.A. Black (ed.), Methods of Soil Analysis, Part 2, Agronomy No. 9. American Society of Agronomy, Madison, WI.

12.5   Hesse, P. R. 1971. A Textbook of Soil Chemical Analysis. Chem. Publishing Co., New York.

12.6   Jones, Jr., J. B. 1973. Soil testing in the United States. Comm. Soil Sci. Plant Anal. 4:307-322.

12.7   Isaac, R. A., and J. D. Kerber. 1971. Atomic absorption and flame photometry: Techniques and uses in soil, plant and water analysis, pp.17-38. IN: L. M. Walsh (ed.) Instrumental Methods for Analysis of Soils and Plant Tissue. Soil Science Society of America, Madison, WI.

12.8   Pratt, P. F.   1965.   Potassium, pp. 1022-1030.   <u>IN</u>: C. A. Black (ed.), Methods of Soil Analysis, Part 2, Agronomy No. 9. American Society of Agronomy, Madison, WI.

12.9   Mehlich, A.   1972.   Uniformity of expressing soil test results: A case for calculating results on a volume basis. Comm. Soil Sci. Plant Anal. 3:417-424.

12.10 Doll, E. C., and R. E. Lucas.   1973.   Testing soils for potassium, calcium and magnesium, pp. 133-152.   <u>IN</u>: L. M. Walsh and J. D. Beaton (eds.), Soil Testing and Plant Analysis, rev. ed. Soil Science Society of America, Madison, WI.

DETERMINATION OF PHOSPHORUS BY MEHLICH NO. 1 (DOUBLE ACID) EXTRACTION

1. PRINCIPLE OF THE METHOD

1.1   This method is primarily used to determine phosphorus in sandy soils which have exchange capacities of less than 10 meq/100 g, are acid in reaction (pH less than 6.5) and are relatively low in organic matter content (less than 5%). The method is not suited for alkaline soils.

1.2   This method was first published by Mehlich (see 12.1) and then by Nelson, Mehlich and Winters (see 12.2) as the North Carolina Double Acid Method. The method, which has been renamed Mehlich No. 1, is adaptable to the coastal plain soils of eastern United States. It is currently being used by a number of state soil testing laboratories in the United States (Alabama, Delaware, Florida, Georgia, Maryland, New Jersey, South Carolina and Virginia) (see 12.3).

2. RANGE AND SENSITIVITY

2.1   Phosphorus can be extracted and determined in soil concentrations from 2-200 kg P/ha without dilution. The upper limit may be extended by diluting the extract prior to colorimetric determination.

2.2   The sensitivity varies depending on the method of color development. Greater sensitivity can be obtained with the molybdophosphoric blue color method (see 12.4) as compared to the vanadomolybdophosphoric acid color method (see 12.4). The estimated precision of the method is $\pm$ 1 mg/L P.

3. INTERFERENCES

3.1   With some soils, the extract may be colored, varying from light to dark yellow. If the vanadomolybdophosphoric acid method (see 12.4) is employed, as originally prescribed for the Mehlich No. 1 acid method (see 12.1), decolorizing is necessary to avoid obtaining high results. Decolorization can be accomplished by including *activated* charcoal in the extraction procedure. Decolorization is not necessary if color development is by the molybdophosphoric blue color procedure (see 12.4). The method is described by Watanabe and Olsen (see 12.5).

3.2   Arsenate present in the extract will produce a blue color with the molybdophosphoric blue color procedure unless the arsenate is reduced. Jackson (see 12.4). describes a reduction procedure.

4. PRECISION AND ACCURACY

4.1   Repeated analyses of two standard soil samples over 30 days in the Georgia Soil Testing and Plant Analysis Laboratory gave coefficients of variation of 6.4% and 9.0%, respectively. The mean value for each soil was 32 and 40 kg/ha, respectively. The variance is essentially a factor related to the heterogeneity of the soil rather than to the extraction or colorimetric procedures.

5. APPARATUS

5.1   No. 10 (2-mm opening) sieve.

5.2   4-cm^3 scoop, volumetric.

5.3   50-mL extraction bottle or flask, with stopper.

5.4   Mechanical reciprocating shaker, 180 oscillations per minute.

5.5   Filter funnel, 11 cm.

5.6   Whatman No. 1 filter paper (or equivalent), 11 cm.

5.7   Photoelectric colorimeter suitable for measurement in the 880 nm range.

5.8   Colorimetric tube or cuvet.

5.9   Funnel racks.

5.10   Analytical balance.

5.11   Volumetric flasks and pipettes as required for preparation of reagents, standard solutions and color development.

6. REAGENTS

6.1   Extracting Reagent (0.05N HCl in 0.025N H_2SO_4) - Pipette 4 mL conc hydrochloric acid (HCl) and 0.7 mL conc sulfuric acid (H_2SO_4) into a 1-liter volumetric flask and dilute to the mark with pure water.

6.2   Ascorbic Acid Solution - Dissolve 176.0 g ascorbic acid in pure water and dilute to 2 liters with pure water. Store in a dark glass bottle in a refrigerated compartment.

6.3   Sulfuric-Molybdate Solution - Dissolve 100 g ammonium molybdate [$(NH_4) Mo_7O_{24} \cdot 4H_2O$] in 500 mL of pure water. Dissolve 2.425 g antimony potassium tartrate [$K(SbO)$ $C_4H_4O_6 \cdot 1/2 H_2O$] in the molybdate solution. Add slowly 1400 mL conc sulfuric acid (H_2SO_4) and mix well. Let it cool and dilute to 2 liters with pure water. Store in a polyethylene or pyrex bottle in a dark, refrigerated compartment.

6.4   Working Solution - Dilute 10 mL ascorbic acid solution (see 6.2) plus 20 mL sulfuric-molybdate solution (see 6.3) with pure water to 1 liter. Allow it to stand at least 1 hour before using. The solution is stable for two to three days.

6.5   Phosphorus Standard (1000 mg/L) - Weigh 3.85 g ammonium dihydrogen phosphate ($NH_4H_2PO_4$) into a 1-liter volumetric flask and bring to volume with Extracting Reagent (see 6.1). Prepare standards containing 1, 2, 5, 10, 15 and 20 mg/L P diluting aliquots of the 1000 mg/L P Standard with Extracting Reagent (see 6.1).

7. PROCEDURE

7.1   Extraction - Weigh 5 g or scoop 4-cm^3 of air-dry, <10-mesh (2-mm) soil into a 50-mL extraction bottle (see 5.1). Add 25 mL Extracting Reagent (see 6.2) and shake for 5 minutes on a reciprocating shaker at a minimum of 180 oscillations per minute (see 5.2). Filter and collect the extract.

7.2   Color Development - Pipette 2 mL extractant into a spectrophotometer cuvet. Add 23 mL Working Solution (see 6.4). Mix well and let it stand for 20 minutes. Read the absorbance at 880 nm with the spectrophotometer. The spectrophotometer should be zeroed against a blank consisting of the Extraction Reagent (see 6.1).

8. CALIBRATION AND STANDARDS

8.1   Working Phosphorus Standards - With the Standard P Solution (see 6.5), prepare 6 working standard solutions containing from 1 to 10 mg/L P in the final volume. Make all dilutions with the Extracting Reagent (see 6.1). Use a 2.0-mL aliquot of each standard and carry through the color development (see 7.2).

8.2   Calibration Curve - On semi-log graph paper, plot absorbance on the logarithmic scale versus mg/L P on the linear scale.

8.3   The color intensity reaches a maximum in approximately 20 minutes and will remain constant for about 6-8 hours.

9. CALCULATION

9.1   When the sample is weighed, mg P/L in filtrate times 10 equals 1 mg P/ha. When the sample is scooped, the results are reported as kg P/ha for a 20-cm depth of soil. If the extraction filtrate is diluted, the dilution factor must be applied.

10. EFFECTS OF STORAGE

10.1   Soils may be stored in an air-dry condition for several months with no effect on extractable phosphorus.

10.2   After extraction, the extraction solution containing phosphorus should not be stored more than 24 hours.

11. INTERPRETATION

11.1   Accurate fertilizer recommendations for phosphorus must be based on known field responses based on local soil-climate-crop conditions (see 12.6). For most soils and crops, the amount of phosphorus extracted is to be interpreted as follows:

Category	kg P/ha (lb/A) in soil	
very low	<11	(<10)
low	11-33	(10-30)
medium	34-67	(31-60)
high	68-112	(61-100)
very high	>112	(>100)

11.2   Interpretations may vary somewhat, depending on soil characteristics and different crops. Interpretative data is given in 12.7.

12. REFERENCES

12.1.   Mehlich, A. 1953. Determination of P, Ca, Mg, K, Na and NH_4. North Carolina Soil Test Division mimeo.

12.2.   Nelson, W. L., A. Mehlich, and E. Winters. 1953. The development, evaluation and use of soil tests for phosphorus availability, pp. 153-188. IN: W. H. Pierre and A. G. Norman (eds.), Soil and Fertilizer Phosphorus. Agronomy No. 4., American Society of Agronomy, Madison, WI.

12.3.   Jones, Jr., J. B. 1973. Soil testing in the United States. Comm. Soil Sci. Plant Anal. 4:307-322.

12.4. Jackson, M. L. 1958. Soil Chemical Analysis. Prentice-Hall, Inc., Englewood Cliffs, NJ.

12.5 Watanabe, F. S., and S. R. Olsen. 1965. Test of an ascorbic acid method for determining phosphorus in water and $NaHCO_3$ extracts from soil. Soil Sc. Soc. Am. Proc. 29:677.

12.6 Isaac, R. A. (ed.) 1983. Reference soil test methods for the southern region of the United States. Southern Coop. Series Bull. 289.

12.7 Thomas, G. W., and D. E. Peaslee. 1973. Testing soil for phosphorus, pp. 115-132. IN: M. Walsh and J. D. Beaton (eds.), Soil Testing and Plant Analysis, rev. ed. Soil Science Society of America., Madison, WI.

DETERMINATION OF POTASSIUM, CALCIUM, MAGNESIUM AND SODIUM BY MEHLICH NO. 1 (DOUBLE ACID) EXTRACTION

1. PRINCIPLE OF THE METHOD

1.1   This method is primarily used to determine potassium, calcium, magnesium and sodium in soils which have exchange capacities of less than 10 meq/100 g, are acid in reaction (pH less than 6.5), and are relatively low in organic matter content (less than 5%). The method is not suited for alkaline soils.

1.2   The use of Mehlich No. 1 as an extraction reagent was first published by Mehlich (see 12.1) and then published specifically as a phosphorus extraction reagent by Nelson, Mehlich and Winters (see 12.2) as the North Carolina Double Acid (now Mehlich No. 1) method. It is adaptable to the coastal plain soils of eastern United States. It is currently being used by a number of state soil testing laboratories in the United States (Alabama, Delaware, Florida, Georgia, Maryland, New Jersey, South Carolina and Virginia) (see 12.3).

2. RANGE AND SENSITIVITY

2.1   Potassium, calcium, magnesium and sodium can be extracted and determined in soil concentrations from 50 to 400 K, 120 to 1200 Ca and 40 to 360 Mg kg/ha without dilution. The range and upper limits may be extended by diluting the extracting filtrate prior to analysis.

2.2   The sensitivity will vary with the type of instrument used, wavelength selected and method of excitation.

2.3   The commonly used methods of analysis are flame emission and atomic absorption spectrometry. For a more complete description of these methods, see Isaac and Kerber (12.4) and Soltanpour et al. (12.12). The use of an AutoAnalyzer for the analysis is described by Flannery and Markus (see 12.5) and Isaac and Jones (see 12.6).

3. INTERFERENCES

3.1   Known interferences and compensation for the changing characteristics of the extract to be analyzed must be acknowledged. The use of internal standards such as lithium and compensating elements such as lanthanum or strontium are essential in most flame methods of excitation (see 12.4 and 12.5).

4 . PRECISION AND ACCURACY

4.1   Repeated analysis of the same soil with medium concentration ranges of potassium, calcium, magnesium and sodium will produce coefficients of variation from 5% to 10%. A major portion of the variance is related to the heterogeneity of the soil rather than to the extraction or method of analysis.

4.2   The level of exchangeable potassium will increase upon the air drying of some soils (see 12.7). Soil samples can be extracted in the moist state. However, the difficulties inherent in handling and storing moist soil hinder the easy adaptation of this method to a routine method of analysis. Compensation can be made based on the expected release of potassium by the particular soil being tested.

5 . APPARATUS

5.1   No. 10 (2-mm opening) sieve.

5.2 5-cm^3 scoop, volumetric.

5.3   50-mL extraction bottle or flask, with stoppers.

5.4   Mechanical reciprocating shaker, 180 oscillations per minute.

5.5   Filter funnel, 11 cm.

5.6   Whatman No. 1 filter paper (or equivalent), 11 cm.

5.7   Flame emission, atomic absorption ICP emission spectrophotometer and/or AutoAnalyzer.

5.8   Funnel racks.

5.9   Analytical balance.

5.10 Volumetric flasks and pipettes as required for preparation of reagents and standard solutions.

6 . REAGENTS

6.1   Extracting Reagent - Pipette 4 mL conc hydrochloric acid (HCl) and 0.7 mL conc sulfuric acid (H_2SO_4) into a 1-liter volumetric flask and dilute to the mark with pure water.

6.2   Potassium Standard (1000 mg/L) - Weigh 1.9080 g potassium chloride (KCl) into a 1-liter volumetric flask and bring to volume with Extraction Reagent (see 6.1). Prepare working standards by diluting aliquots of the stock solution standard with Extraction Reagent (see 6.1) to cover the anticipated range in concentration to be found in the soil extraction filtrate. Working standards from 5 to 50 mg K/L should be sufficient for most soils.

6.3   Calcium Standard (1000 mg/L) - Weigh 2.498 g calcium carbonate (CaCO$_3$) into a 1-liter volumetric flask, add 50 mL pure water and add dropwise a minimum volume (approximately 20 mL) of conc hydrochloric acid (HCl) to effect the complete solution of the calcium carbonate. Dilute to the mark with Extraction Reagent (see 6.1). Prepare working standards by diluting aliquots of the stock solution standard with Extraction Reagent (see 6.1) to cover the anticipated range in concentration to be found in the soil extraction filtrate. Working standards from 15 to 150 mg Ca/L should be sufficient for most soils.

6.4   Magnesium Standard (1000 mg/L) - Weigh 1.000 g magnesium ribbon into a 1-liter volumetric flask and dissolve in the minimum volume of dilute hydrochloric acid (HCl) and dilute to the mark with Extraction Reagent (see 6.1). Prepare working standards by diluting aliquots of the stock solution standard with Extraction Reagent (see 6.1) to cover the anticipated range in concentration to be found in the soil extraction filtrate. Working standards from 5 to 50 mg Mg/L should be sufficient for most soils.

6.5   Sodium Standard (1000 mg/L - Weigh 2.542 g sodium chloride (NaCl) into a 1-liter volumetric flask and bring to volume with Extraction Reagent (see 6.1). Prepare working standards by diluting aliquots of the stock solution standard with Extraction Reagent (see 6.1) to cover the anticipated range in concentration to be found in the soil extraction filtrate. Working standards from 1 to 10 mg Na/L should be sufficient.

7. PROCEDURE

7.1   Extraction - Weigh 5 g or scoop 5-cm^3 (see 5.2) air-dry, <10-mesh (2-mm) soil into a 50-mL extraction bottle (see 5.3). Add 25 mL of the Extraction Reagent (see 6.1) and shake for 5 minutes on a reciprocating shaker (see 5.4). Filter and collect the filtrate.

7.2   Analysis - The elements potassium, calcium, magnesium and sodium in the filtrate can be determined by either flame emission or atomic absorption ICP emission spectrometry, or by using an AutoAnalyzer. Since instruments vary in their operating conditions, no specific details are given here. It is recommended that the procedures described by Isaac and Kerber (see 12.4), Soltanpour et al (12.12), Flannery and Markus (12.5), or Isaac and Jones (12.6) be followed, depending on the technique used.

8. CALIBRATION AND STANDARDS

8.1   Working Standards - Working standards should be prepared as described in section 6. If element concentrations are found outside the range of the instument or standards, suitable dilutions should be prepared starting with a 1:1 soil extract to Extraction Reagent (see 6.1) dilution. Dilution should be made so as to minimize the magnification of errors introduced by diluting.

8.2   Calibration - Calibration procedures vary with instrument techniques and the type of instrument. Every precaution should be taken to ensure that the proper procedures are followed and that the manufacturer's recommendations are followed in the operation and calibration of the instrument.

9. CALCULATION

9.1   The results are reported as kg/ha for a 20-cm depth of soil. Kg element/ha = mg/L element in extraction filtrate x 10. If the extraction filtrate is diluted, the dilution factor must be applied.

9.2   To convert to other units for comparisons, see Mehlich (12.8).

10. EFFECTS OF STORAGE

10.1   Soils may be stored in an air-dry condition for several months with no effect on the exchangeable potassium, calcium, magnesium and sodium content.

10.2   After extraction, the filtrate containing potassium, calcium, magnesium and sodium should not be stored longer than 24 hours unless it is refrigerated or treated to prevent bacterial growth.

11. INTERPRETATION

11.1   An evaluation of the analytical results for the determination of fertilizer recommendations, particularly for the elements potassium and magnesium, must be based on field response data conducted under local soil-climate-crop conditions (see 12.9). Interpretative data used in the Southeast is available (see 12.10).

12. REFERENCES

12.1. Mehlich, A. 1953. Determination of P, Ca, Mg, K, Na and NH_4 (mimeo). North Carolina Soil Test Division, Raleigh, NC.

12.2. Nelson, W. L., A. Mehlich, and E. Winters. 1953. The developments, evaluation and use of soil tests for phosphorus availability, pp. 153-188. IN: W. H. Pierre and A. G. Norman (eds.), Soil and Fertilizer Phosphorus. Agronomy No. 4 American Society of Agronomy, Madison, WI.

12.3. Jones, Jr., J. B. 1973. Soil testing in the United States. Comm. Soil Sci. Plant Anal. 4:307-322.

12.4. Isaac, R. A., and J. D. Kerber. 1972. Atomic absorption and flame photometry: Techniques and uses in soil, plant and water analyses, pp. 17-38. IN: L. M. Walsh (ed), Instrumental Methods for Analysis of Soils and Plant Tissue. Soil Science Society of America, Madison, WI.

12.5 Flannery, R. L., and D. K. Markus. 1972. Determination of phosphorus, potassium, calcium and magnesium simultaneously in North Carolina, ammonium acetate, and Bray Pl soil extracts by AutoAnalyzer, pp. 97-112. IN: L. M. Walsh (ed.) Instrumental Methods for Analysis of Soils and Plant Tissue. Soil Science Society of America, Madison, WI.

12.6 Isaac, R. A., and J. B. Jones, Jr. 1971. AutoAnalyzer systems for the analysis of soil and plant tissue extracts, pp. 57-64. IN: Technicon International Congress, Terrytown, NY.

12.7 Pratt, P. F. 1965. Potassium, pp. 1022-1030. IN: C. A. Black (ed.), Methods of Soil Analysis, Part 2. Agronomy No. 9. American Society of Agronomy, Madison, WI.

12.8 Mehlich, A. 1972. Uniformity of expressing soil test results. A case for calculating results on a volume basis. Comm. Soil. Sci. Plant Anal. 3:417-424.

12.9 Doll, E. C., and R. E. Lucas. 1973. Testing soils for potassium, calcium and magnesium, pp. 133-152. IN: L. M. Walsh and J. D. Beaton (eds.), Soil Testing and Plant Analysis. Soil Science Society of America, Madison, WI.

12.10 Jones, Jr., J. B., J. T. Cope,and J. D. Lancaster. 1974. Soil testing methods used by the 13 southeastern states. Southern Cooperative Bulletin 190. Georgia Cooperative Extension Service, Athens, GA.

12.11 Mehlich, A. 1974. Uniformity of soil test results as influenced by extractants and soil properties, pp. 295-305. IN: J. Wehrmann (ed.), Proceedings 7th International Colloquium Plant Analysis and Fertilizer Problems, Hanover, Germany.

12.12 Soltanpour, P. N., J. B. Jones, Jr., and S. M. Workman. 1982. Optical emission spectrometry, pp. 29-65. IN: A. L. Page et al. (eds.), Methods of Analysis, Part 2, rev. ed. American Society of Agronomy, Madison, WI.

DETERMINATION OF ZINC BY MEHLICH NO. 1 (DOUBLE ACID) EXTRACTION

1. PRINCIPLE OF THE METHOD

1.1   This method for determining extractable zinc has been evaluated only on soils which have cation-exchange capacities of less than 10 meq/100 gm, are acid in reaction (pH less than 7.0), and are relatively low in organic matter content (less than 5%). Its suitability for use on alkaline or organic soils has not been determined. This method is described in some detail by Perkins (see 12.1) for use with the sandy coastal plain soils of the southeastern United States.

1.2   The use of Mehlich No. 1 as an extraction reagent for cations and phosphorus was first reported by Mehlich (see 12.2), and later classified as a phosphorus extraction reagent by Nelson, Mehlich and Winters (see 12.3) as the North Carolina Double Acid Method. Wear and Evans (see 12.4) compared Mehlich No. 1 versus 0.1N HCl and EDTA as a zinc extraction reagent on 12 soils and found Mehlich No. 1 extractable zinc to corre-late more closely with zinc uptake by corn and sorghum plants. Alley et al. (see 12.5) developed a prediction equation for zinc response under field conditions, using Mehlich No. 1 extractable zinc. The equation was improved considerably by taking into consideration soil pH and Mehlich No. 1 extractable phosphorus.

2. RANGE AND SENSITIVITY

2.1   Zinc can be extracted and determined in soil concentrations from 0.4 to 8.0 mg Zn/L without dilution. The range and upper limits may be extended by diluting the extracting filtrate prior to analyses.

2.2   Sensitivity will vary with the type of instrument used, wavelength and method of excitation.

2.3   The commonly used method of analysis is atomic absorption or ICP emission spectrometry. Isaac and Kerber (see 12.6) provide a more complete description of the former method; Soltanpour (see 12.10) describes the latter method.

3. INTERFERENCES

3.1   Known interferences and compensation for changing characteristics of the extract to be analyzed must be acknowledged. However, no serious interferences have been reported.

4. PRECISION AND ACCURACY

4.1   The major source of variance in the extraction is the heterogeneity of the soil sample itself. Repeated analyses of the same soil sample will give a coefficient of variation of approximately 10%.

5. APPARATUS

5.1   No. 10 (2-mm opening) sieve.

5.2   4-cm^3 scoop, volumetric.

5.3   50-mL extraction bottle or flask, with stoppers.

5.4   Mechanical reciprocating shaker, 180 oscillations per minute.

5.5   Filter funnel, 11 cm.

5.6   Acid-washed filter paper, Whatman No. 42, 11 cm or equivalent.

5.7   Atomic absorption or ICP emission spectrophotometer.

5.8   Funnel racks.

5.9   Analytical balance.

5.10 Volumetric flasks and pipettes as required for preparation of reagents and standard solutions.

6. REAGENTS

6.1   Extraction Reagent - Pipette 4 mL conc hydrochloric acid (HCl) and 0.7 mL conc sulfuric acid (H_2SO_4) into a 1-liter volumetric flask and dilute to the mark with pure water.

6.2   Zinc Standard (1000 mg/L) - Dissolve 4.3478 g zinc sulfate ($ZnSO_4 \cdot 7H_2O$) in 1 liter of Extraction Reagent (see 6.1). Prepare working standards by diluting aliquots of the stock solution standard with Extraction Reagent (see 6.1) to cover the anticipated range in concentration to be found in the soil extraction filtrate. Working standards from 0 to 3 mg Zn/L in solution should be sufficient for most soils.

7. PROCEDURE

7.1   Extraction - Weigh 5 g or scoop 4-cm^3 air-dry <10-mesh (2-mm) soil into an acid-washed 50-mL extraction bottle. Add

20 mL of Extraction Reagent (see 6.1) and shake for 5 minutes on a reciprocating shaker (see 5.4). Filter and collect the filtrate.

7.2 <u>Analysis</u> - Since atomic absorption and ICP emission spectrometer instruments vary in operating conditions, no specific details are provided here.

8. CALIBRATION AND STANDARDS

8.1 <u>Working Standards</u> - Working standards should be prepared as described in section 6. If concentrations are found outside the range of the instrument or standards, suitable dilutions should be prepared starting with a 1:1 soil-to-Extraction Reagent dilution. Dilution should be made so as to minimize the magnification of errors introduced by diluting.

8.2 <u>Calibration</u> - Calibration procedures vary with instrument techniques and the type of instrument. Every precaution should be taken to ensure that the proper procedures are followed and that the manufacturer's recommendations in the operation and calibration of the instrument are used.

9. CALCULATION

9.1 The results can be reported as lb/A of extractable zinc in the soil, multiplying the mg Zn/L in the filtrate x 8. To convert to other units for comparison, see Mehlich (see 12.7).

10. EFFECTS OF STORAGE

10.1 Soils may be stored in an air-dried condition for several months with no effects on the extractable zinc content.

10.2 After extraction, the filtrate should not be stored longer than 24 hours without refrigeration or other treatment to prevent bacterial growth.

11. INTERPRETATION

11.1 An evaluation of the results as well as zinc recommendations must be based on field response data conducted under local soil-climate-crop conditions (see 12.8). Interpretative data that would be applicable to the Southeast are given by Alley et al. (see 12.5), Perkins (see 12.1), and Cox and Wear (see 12.9).

11.2 The critical soil test zinc level for corn as interpreted by Cox and Wear (see 12.9) is 0.8 mg Zn/L. The probability of a corn yield response to zinc fertilization on soils testing below this value would be high. This critical level may not apply to extremely high CEC soils, high P soils or very acid soils.

12. REFERENCES

12.1. Perkins, H. F. 1970. A rapid method of evaluating the zinc status of coastal plain soils. Comm. Soil Sci. Plant Anal. 1:35-42.

12.2. Mehlich, A. 1953. Determination of P, Ca, Mg, K, Na and NH_4 (mimeo). North Carolina Soil Testing Division, Raleigh, NC.

12.3. Nelson, W. L., A. Mehlich, and E. Winters. 1953. The development, evaluation and use of soil tests for phosphorus availability, pp. 153-188. IN: W. H. Pierre and A. G. Norman (eds.), Soil and Fertilizer Phosphorus, Agronomy No. 4. American Society of Agronomy, Madison, WI.

12.4. Wear, John I., and Clyde E. Evans. 1968. Relationships of zinc uptake of corn and sorghum to soil zinc measured by three extractants. Soil Sci. Soc. Am. Proc. 32:543-546.

12.5 Alley, M. M., D. C. Martens, M. G. Schnappinger, and G. W. Hawkins. 1972. Field calibration of soil tests for available zinc. Soil Sci. Soc. Am. Proc. 36:621-624.

12.6 Isaac, R. A., and J. D. Kerber. 1972. Atomic absorption and flame photometry: Techniques and uses in soil, plant and water analysis, pp. 17-38. IN: L. M. Walsh (ed.), Instrumental Methods of Analysis of Soils and Plant Tissue. Soil Science Society of America, Madison, WI.

12.7 Mehlich, A. 1972. Uniformity of expressing soil test results: A case for calculating results on a volume basis. Comm. Soil Sci. Plant Anal. 3:417-424.

12.8 Viets, Jr., F. G., and W. L. Lindsay. 1973. Testing soils for zinc, copper, manganese and iron, pp. 153-172. IN: L. M. Walsh and J. D. Beaton (eds.), Soil Testing and Plant Analysis, rev. ed. Soil Science Society of America, Madison, WI.

12.9 Cox, F. R., and J. I. Wear (eds.). 1977. Diagnosis and correlation of zinc problems in corn and rice production. Southern Cooperative Series Bulletin 222. North Carolina State Experiment Station, Raleigh, NC.

12.10 Soltanpour, P. N., J. B. Jones, Jr., and S. M. Workman. 1982. Optical emission spectrometry, pp. 29-65. IN: A.L. Page et al. (eds.), Methods of Soil Analysis, Part 2, rev. ed. Soil Science Society of America, Madison, WI.

DETERMINATION OF PHOSPHORUS BY MORGAN EXTRACTION

1. PRINCIPLE OF THE METHOD

1.1   This method is primarily for determining phosphorus content in acid soils with cation-exchange capacities of less than 20 meq/100 g.

1.2   This method, first proposed by Morgan (see 12.1), was described in detail by Lunt et al. (see 12.1) and later by Greweling and Peech (see 12.3). The extracting reagent is well buffered at pH 4.8, and, when used in conjunction with activated carbon, yields clear and colorless extracts. The Morgan method was used by several state soil testing laboratories in the northeastern and northwestern United States (see 12.4).

2. RANGE AND SENSITIVITY

2.1   Phosphorus can be extracted and determined in soil concentrations from 2 to 100 kg P/ha without dilution. The upper limit may be extended by diluting the extract prior to colorimetric determination.

2.2   The sensitivity varies depending on the method of color development. Greater sensitivity can be obtained with the molybdophosphoric blue color method (see 12.5) as compared to the vanado-molybdophosphoric acid color method (see 12.5). The estimated precision of the method is $\pm$ 1 ppm P.

3. INTERFERENCES

3.1   With some soils, the extract may be colored, varying from light to dark yellow. If the vanado-molybdophosphoric acid method (see 12.5) is employed, decolorizing is necessary to avoid obtaining high results. Decolorization can be accomplished by including activated charcoal in the extraction procedure. Decolorization is not necessary if color development is by the molybdophosphoric blue color procedure (see 12.5 and 12.6). A description of the method is given by Watanabe and Olsen (see 12.7).

3.2   Arsenate present in the extractant will produce a blue color with the molybdophosphoric blue color procedure unless the arsenate is reduced. A reduction procedure is given by Jackson (see 12.5).

4. PRECISION AND ACCURACY

4.1 Repeated analyses of two standard soil samples over 30 days in the Georgia Soil Testing and Plant Analysis Laboratory gave coefficients of variability of 6.4 % to 9.0%, respectively. Each soil tested 32 and 40 kg P/ha, respectively. The variance factor is essentially related to the homogeneity of the soil rather than to the extraction or colorimetric procedures.

5. APPARATUS

5.1 No. 10 (2-mm opening) sieve.

5.2 5-cm^3 scoop, volumetric.

5.3 Extraction bottle or flask, 50-mL w/stoppers.

5.4 Mechanical reciprocating shaker, 180 oscillations/minute.

5.5 Filter funnel, 11 cm.

5.6 Whatman No. 1 filter paper (or equivalent), 11 cm.

5.7 Photoelectric colorimeter suitable for measurement in the 880 nm range.

5.8 Colorimetric tube or cuvet.

5.9 Funnel racks.

5.10 Analytical balance.

5.11 Volumetric flasks and pipettes as required for the preparation of reagents, standard solutions and color development.

6. REAGENTS

6.1 Extracting Reagent - Dissolve 100 g sodium acetate ($NaC_2H_3O_2 \cdot 3H_2O$) in about 900 mL pure water. Add 30 mL glacial acetic acid ($HC_2H_3O_2$), adjust the pH to 4.8, and dilute to 1 liter with pure water.

6.2 Ascorbic Acid Solution - Dissolve 176.0 g ascorbic acid in pure water and dilute to 2 liters with pure water. Store in a dark glass bottle in a refrigerated compartment.

6.3   Sulfuric-Molybdate Reagent - Dissolve 100 g ammonium molybdate [$(NH_4)_6MO_7O_{24} \cdot 4H_2O$] in 500 mL of pure water. Dissolve 2.425 g antimony potassium tartrate [$K(SbO)C_4H_4P_6$ $1/2H_2O$] in the molybdate solution. Add slowly 1400 mL conc. H_2SO_4 and mix well. Let cool and dilute to 2 liters with pure water. Store in a polyethylene or pyrex bottle in a dark refrigerated compartment.

6.4   Working Solution - Dilute 10 mL Ascorbic Acid Solution (see 6.3) plus 20 mL Sulfuric-Molybdate Reagent (see 6.3) with Extracting Reagent (see 6.1) to 1 liter. Prepare fresh daily. Allow it to stand at least 2 hours before using.

6.5   Phosphorus Standard (100 ppm P) - Weigh 0.4394 g monobasic potassium phosphate (KH_2PO_4) which has been overdried at $100°$ C into a 1-liter volumetric flask. Bring to volume with Extracting Reagent (see 6.1).

7. PROCEDURE

7.1   Extraction - Measure 5 cm^3 of air-dry, <10-mesh (2-mm) soil into a 50-mL extraction bottle (see 5.3). Add 25 mL of the Extracting Reagent (see 6.1) and shake for 5 minutes on a reciprocating shaker at a minimum of 180 oscillations per minute (see 5.4). Filter and collect the extract.

7.2   Color Development - Pipette 1 mL extractant into a spectro-photometer cuvet. Add 24 mL Working Solution (see 6.4). Mix well and let stand 20 minutes. Read the absorbance at 882 nm. The spectrometer should be zeroed against a blank consisting of 1 mL Extracting Reagent (see 6.1) and 24 mL Working Solution (see 6.4).

8. CALIBRATION AND STANDARDS

8.1   Working Phosphorus Standards - With the Standard Phosphorus Solution (see 6.5), prepare 6 working standard solutions containing from 1 to 20 ppm phosphorus in the final volume. Make all dilutions with the Extracting Reagent (see 6.1). Use a 1.0-mL aliquot of each standard and carry through the color development (see 7.2).

8.2   Calibration Curve - On semi-log graph paper, plot the percentage of transmittance on the logarithmic scale versus ppm phosphorus in the standards on the linear scale.

8.3   The color intensity reaches a maximum in approximately 20 minutes and will remain constant for about 24 hours.

9. CALCULATION

9.1 The results are reported as kg P/ha for a 20 cm depth of soil. Kg P/ha = ppm in extractant x 10.

10. EFFECTS OF STORAGE

10.1 Soils may be stored in an air-dry condition for several months with no effect on extractable P.

10.2 After extraction, the extraction solution containing phosphorus should not be stored for more than 24 hours.

11. INTERPRETATION

11.1 Accurate fertilizer recommendations for phosphorus must be based on field response data conducted under local soil-climate-crop conditions (see 12.7 and 12.8).

11.2 Interpretations may vary somewhat, depending on soil characteristics and different crops.

12. REFERENCES

12.1 Morgan, M. F. 1941. Chemical soil diagnosis by the Universal Soil Testing System. Conn. Agr. Exp. Sta. (New Haven) Bull. 450.

12.2 Lunt, H. A., C. L. W. Swanson, and H. G. M. Jacobson. 1950. The Morgan Soil Testing System. Conn. Agr. Exp. Sta. (New Haven) Sta. Bull. 541.

12.3 Greweling, T., and M. Peech. 1965. Chemical soil tests. Cornell Agr. Exp. Sta. Bull. 960.

12.4 Jones, Jr., J. B. 1973. Soil testing in the United States. Comm. Soil Sci. Plant Anal. 4:307-322.

12.5 Jackson, M. L. 1958. Soil Chemical Analysis. Prentice-Hall, Inc., Englewood Cliffs, NJ.

12.6 Olsen, S. R., and L. E. Sommers. 1982. Phosphorus, pp. 403-430. IN: A. L. Page (ed.), Methods of Soil Analysis, Part 2. Chemical and Microbiological Properties, 2nd ed. American Society of Agronomy, Madison, WI.

12.7 Watanabe, F. S., and S. R. Olsen. 1965. Test of an ascorbic acid method for determining phosphorus in water and $NaHCO_3$ extracts from soil. Soil Sci. Soc. Amer. Proc. 29:677-678.

12.8   Thomas, G. W., and D. E. Peaslee. 1973. Testing soil for phosphorus, pp. 115-132. <u>IN</u>: L. M. Walsh and J. D. Beaton (eds.), Soil Testing and Plant Analysis, rev. ed. Soil Science Society of America, Madison, WI.

DETERMINATION OF POTASSIUM, CALCIUM AND MAGNESIUM BY MORGAN EXTRACTION

1. PRINCIPLE OF THE METHOD

1.1 This method is primarily used for determining potassium, calcium and magnesium in acid soils with cation exchange capacities of less than 20 meq/100 g.

1.2 This method, proposed by Morgan (see 12.1), was described in detail first by Lunt et al. (see 12.2) and later by Greweling and Peech (see 12.3). The Morgan method was used by a number of soil testing laboratories in the northeastern and northwestern United States (see 12.4). The Extracting Reagent is well buffered at pH 4.8 and, when used in conjunction with activated carbon, yields clear and colorless extracts. The concentration of sodium acetate is sufficiently high to effect replacement of about 80% of the exchangeable cations.

2. RANGE AND SENSITIVITY

2.1 The range of detection will depend upon the setup of the particular instrument. The range can be extended by diluting the extract prior to colorimetric determination.

2.2 The sensitivity will vary with the type of instrument used, wave length selected and method of excitation.

2.3 The commonly used methods of analysis are flame emission and atomic absorption spectroscopy. A more complete description of these methods is given by Isaac and Kerber (see 12.5).

3. INTERFERENCES

3.1 Under certain conditions, the Extracting Reagent (see 6.1) will extract more than those elements which exist in the soil only in an exchangeable form, such as, in the soil solution, those elements released by weathering action during the period of extraction, and the dissolution of carbonates of calcium and magnesium if these are present in the soil. However, these contributions will not normally significantly alter the analysis when it is used to assess the fertility status of the soil.

3.2 Known interferences and compensation for the changing characteristics of the extract to be analyzed must be acknowledged. The use of an internal standard such as lithium and a compensating element such as lanthanum is essential in most flame methods of excitation (see 12.5).

4. PRECISION AND ACCURACY

4.1   Repeated analysis of the same soil with medium concentration ranges of potassium, calcium and magnesium will give coefficients of variability of from 5% to 10%. A major portion of the variance is related to the heterogeneity of the soil rather than to the extraction or method of analysis.

4.2   The level of exchangeable potassium will increase upon the air drying of some soils (see 12.6). Soil samples can be extracted in the moist state. However, the difficulties inherent in handling and storing moist soil makes this method difficult for easy adaptation to a routine method of analysis. Compensation can be made based on the expected release of potassium by the particular soil being tested.

5. APPARATUS

5.1   No. 10 (2-mm opening) sieve.

5.2   5 cm^3 scoop, volumetric.

5.3   Extraction bottle or flask, 50-mL, w/stoppers.

5.4   Mechanical reciprocating shaker, 180 oscillations/minute.

5.5   Filter funnel, 11 cm.

5.6   Whatman No. 1 filter paper (or equivalent), 11 cm.

5.7   Flame Emission and /or Atomic Absorption Spectrophotometer.

5.8   Funnel racks.

5.9   Analytical balance.

5.10   Volumetric flasks and pipettes as required for preparation of reagents and standard solutions.

6. REAGENTS

6.1   <u>Extracting Reagent</u> - Dissolve 100 g sodium acetate ($NaC_2H_3O_2 \cdot 3H_2O$) in about 900 mL pure water. Add 30 mL glacial acetic acid ($HC_2H_3O_2$), adjust the pH to 4.8, and dilute to 1 liter with pure water.

6.2 Potassium Standard (1000 mg/L) - Weigh 1.9080 g potassium chloride (KCl) into a 1-liter volumetric flask and bring to volume with Extracting Reagent (see 6.1). Prepare working standards by diluting aliquots of the stock solution standard with the Extracting Reagent to cover the anticipated range in concentration to be found in the soil extraction filtrate. Working standards from 5 to 100 mg K/L should be sufficient for most soils.

6.3 Calcium Standard (1000 mg/L) - Weigh 2.498 g calcium carbonate ($CaCO_3$) into a 1-liter volumetric flask, add 50 mL of pure water, and add dropwise a minimum volume of conc. hydrochloric acid (HCl) (approximately 20 mL) to effect complete solution of the calcium carbonate. Dilute to the mark with Extracting Reagent (see 6.1). Prepare working standards by diluting aliquots of the stock solution standard with Extracting Reagent to cover the anticipated range in concentration to be found in the soil extraction filtrate. Working standards from 50 to 220 mg Ca/L should be sufficient for most soils.

6.4 Magnesium Standard (1000 mg/L) - Weigh 1.000 g magnesium ribbon into a 1-liter volumetric flask and dissolve in a minimum volume of (1+1) HCl and dilute in one liter of Extracting Reagent (see 6.1). Prepare working standards by diluting aliquots of the stock solution standard with Extracting Reagent to cover the anticipated range in concentration to be found in the soil extraction filtrate. Working standards from 5 to 50 mg Mg/L should be sufficient for most soils.

7. PROCEDURE

7.1 Extraction - Measure 5 cm^3 (see 5.2) of air-dry, <10-mesh (2-mm) soil into a 50-mL extraction bottle (see 5.3). Add 25 mL Extracting Reagent (see 6.1) and shake for 5 minutes on a reciprocating shaker at a minimum of 180 oscillations per minute (see 5.4). Filter and collect the filtrate.

7.2 Analysis - The elements potassium, calcium and magnesium in the filtrate can be determined by either flame emission or atomic absorption spectroscopy. Since instruments vary in their operating conditions, no specific details are given here. It is recommended that the procedures described by Isaac and Kerber (see 12.5) be followed.

8. CALIBRATION AND STANDARDS

8.1 Working Standards - Working standards should be prepared as described in section 6. If element concentrations are found outside the range of the instrument or standards, suitable dilutions should be prepared starting with a 1:1 soil extract to Extracting Reagent (see 6.1) dilution. Dilution should be made so as to minimize the magnification of errors introduced by diluting.

8.2   Calibration Curve - Calibration procedures vary with instrumental techniques and type of instrument. Every precaution should be taken to ensure that the instrument is properly calibrated and used.

9 . CALCULATION

9.1   The results are reported as kg/ha for a 20 cm depth of soil. Kg of element/ha = mg of element per liter in extraction filtrate x 10.

9.2   To convert to other units for comparison, see Mehlich (12.7).

10. EFFECTS OF STORAGE

10.1   Soils may be stored in an air-dry condition for several months with no effect on the exchangeable potassium, calcium and magnesium content. Potassium may be released on drying for some soils (see 12.6).

10.2   After extraction, the filtrate containing potassium, calcium and magnesium should not be stored longer than 24 hours unless it is refrigerated or treated to prevent bacterial growth.

11. INTERPRETATION

11.1   An evaluation of the analysis results as well as accurate fertilizer recommendations, particularly for the elements potassium and magnesium, must be based on field response data conducted under local soil-climate-crop conditions (see 12.8).

12. REFERENCES

12.1   Morgan, M. F. 1941. Chemical diagnosis by the Universal Soil Testing System. Conn. Agr. Exp. Sta. (New Haven) Bull. 450.

12.2   Lunt, H. A., C. L. W. Swanson, and H. G. M. Jacobson. 1950. The Morgan Soil Testing System. Conn. Agr. Exp. Sta. (New Haven) Bull. 541.

12.3   Greweling, T., and M. Peech. 1965. Chemical Soil Tests. Cornell Agr. Exp. Sta. Bull. 960.

12.4   Jones, Jr., J. B. 1973. Soil testing in the United States. Comm. Soil Sci. Plant Anal. 4:307-322.

12.5   Isaac, R. A., and J. D. Kerber. 1971. Atomic absorption and flame photometry: Techniques and uses in soil, plant and water analysis, pp. 17-38. IN: L. M. Walsh (ed.), Instrumental Methods for Analysis of Soils and Plant Tissue. Soil Science Society of America, Madison, WI.

12.6   Hesse, P. R. 1971. A Textbook of Soil Chemical Analysis. Chemical Publishing Co., New York, NY.

12.7   Mehlich, A. 1972. Uniformity of expressing soil test results: A case for calculating results on a volume basis. Comm. Soil Sci. Plant Anal. 3:417-424.

12.8   Doll, E. C., and R. E. Lucas. 1973. Testing soils for potassium, calcium and magnesium, pp. 133-152. IN: L. M. Walsh and J. D. Beaton (eds.), Soil Testing and Plant Analysis, Soil Science Society of America, Madison, WI.

DETERMINATION OF PHOSPHORUS BY MEHLICH NO. 3 EXTRACTION

1. PRINCIPLE OF THE METHOD

1.1   The extraction of phosphorus by this procedure is designed to be applicable across a wide range of soil properties ranging in reaction from acid to basic. This method correlates well with Bray P1 on acid to neutral soil ($r^2 = 0.966$). It does not correlate with Bray P1 on calcareous soils. The Mehlich No. 3 method correlates with the Olsen extractant on calcareous soils ($r^2 = 0.918$), even though the quantity of Mehlich No. 3 extractable phosphorus is considerably higher.

1.2   This extractant was developed by Dr. Adolf Mehlich in 1981 and described by A. L. Hatfield for the late Dr. Mehlich (see 10.2). This procedure was developed on a 1:10 soil-solution ratio (2.5 cm^3 soil + 25 cm^3 extractant) for a 5-minute shaking period at 200, 4-cm reciprocations/ minute.

2. RANGE AND SENSITIVITY

2.1   Phosphorus can be extracted and determined in soil concentrations 2-400 kg/ha without dilution, using the molybdophosphic blue color procedure first described by Murphy and Riley (see 10.3) and modified by Watanabe and Olsen (see 10.4).

3. PRECISION AND ACCURACY

3.1   Repeated analyses of two standard soil samples for 36 separate extractions by the NCDA Soil Testing Laboratory gave a variance of 6.28% to 6.39%, respectively. Each soil tested 97 and 132 mg P/dm^3, respectively. The variance is most likely related more to the heterogeneity of the samples rather than to measurement, extraction or colorimetric procedures.

4. APPARATUS

4.1   No. 10 (2-mm opening) sieve.

4.2   2.5 cm^3 (volumetric) soil measure and teflon coated leveling rod.

4.3   100-mL extraction bottles, plastic or glass, are suitable.

4.4   Reciprocating shaker (200, 4-cm reciprocations/minute).

4.5 Filter funnels, 11 cm.

4.6 Whatman No. 1 (or equivalent) filter paper, 11 cm.

4.7 Funnel rack.

4.8 Vials, polystyrene plastic, 25- and 50-mL capacity, for collection of extract and color development, respectively.

4.9 Automatic extractant dispenser, 25-mL capacity. Other dispensers or pipettes could be used, depending on preference.

4.10 Dilutor-dispenser apparatus for delivery of sample and color development reagent.

4.11 Volumetric flasks and pipettes are required for preparation of reagents and standard solutions. Pipettes could also be used for color development.

4.12 Photometric colorimeter suitable for measurement in the 880-nm range. Colorimeters equipped with moveable fiber optic probes can be used to read samples directly from the color development vials.

5. REAGENTS

5.1 All reagents are ACS analytical grade unless otherwise stated.

5.2 Extracting Reagent (0.2N CH_3COOH; 0.25N NH_4NO_3; 0.015N NH_4F; 0.13N HNO_3; 0.001M EDTA). For specific procedure on making up extractant, see 10.2.

5.3 Ascorbic Acid Solution - Dissolve 176.0 g ascorbic acid ($C_6H_8O_6$) in pure water and dilute to 2 liters with pure water. Store solution in a dark glass bottle in a refrigerated compartment.

5.4 Sulfuric-Molybdate-Tartrate Solution - Dissolve 100 g ammonium molybdate [$(NH_4)_6 Mo_7O_{24} \cdot 4H_2O$] in 500 mL of pure water. Dissolve 2.425 g antimony potassium tartrate [$K(SbO) C_4H_4O_6 \cdot 1/2 H_2O$] in molybdate solution. Add slowly 1400 mL concentrated H_2SO_4 and mix well. Let it cool and dilute to 2 liters with pure water. Store the solution in a polyethylene or pyrex bottle in a dark, refrigerated compartment.

5.5   Working Solution - Dilute 10 mL of the ascorbic acid solution (see 5.3) plus 20 mL of the sulfuric-molybdate-tartrate solution (see 5.4), with the Extracting Reagent (see 5.2) to make 1 liter. Allow the solution to come to room temperature before using. Prepare fresh daily.

5.6   Phosphorus Standard (200 mg P/L) - Dissolve 0.879 g monopotassium phosphate (KH_2PO_4) in approximately 500 mL of Extracting Reagent (see 5.2) and bring to 1-liter volume with Extracting Reagent (see 5.2). Prepare standards containing 1, 2, 5, 10, 15 and 20 mg P/L by diluting appropriate aliquots of 200 mg P/L with Extracting Reagent (see 5.2).

6. PROCEDURE

6.1   Extraction - Measure 2.5 cm^3 of air-dry, 10-mesh (2-mm) soil into a 100-mL extraction bottle (see 4.3). Add 25 cm^3 of Extracting Reagent (see 5.2) and shake for 5 minutes on a reciprocating shaker (see 4.4). Filter and collect the extract. For the rationale of using a volume soil measure, see Mehlich (10.1).

6.2   Color Development - Using a pipette or dilutor dispenser, dilute 1 mL of the sample extract (see 6.1) with 27 mL of the working solution (see 5.5). Allow the color to develop for at least 20 minutes before reading. Read the transmission at 880 nm (2 cm light path probe colorimeter) or 882 nm for the standard cuvette-type colorimeter.

6.3   The color intensity reaches its maximum in approximately 20 minutes and will remain constant for about 6 hours.

7. CALIBRATION AND STANDARDS

7.1   Working Phosphorus Standards - After the calibration curve is established (see 5.6), the instrument can be calibrated for routine analysis using Extracting Reagent (see 5.2) as the blank and the 20 mg P/L standard (see 5.6). The 20 mg P/L standard should read 10% T following color development at the 1:27 sample-to-working-solution ratio (see 6.2). The blank solution should be diluted at the same 1:27 ratio as the standards. If the instrument reading is significantly above or below 10% T, check the standard preparation, the sample dispenser, or the dilution ratio between the standard and the working solution. The probe colorimeter technique requires a 1:27 sample-to-working-solution ratio to achieve a 10% T reading at 20 mg P/L. A linear curve can be obtained by converting %T to optical density (OD).

Other types of colorimeters may require a different sample-to-working-solution ratio.

8. CALCULATION

8.1 The results are reported as mg P/dm^3 (mg P/L of the standard or phosphorus soil extract multiplied by 10). Multiply the mg P/dm^3 by 2 to obtain the kg P/ha for a 20-cm depth of soil. If the soil extract requires dilution, multiply the results by the appropriate dilution factor.

8.2 To convert soil test values to other units, see Mehlich (10.1).

9. INTERPRETATION

9.1 Critical phosphorus levels proposed by Mehlich (see 10.5) are listed below.

Category	mg P/dm^3	kgP/ha	Expected Crop Response
Very low	<20	<40	definite
Low	20-30	40- 60	probable
Medium	31-50	62-100	less likely
High	>50	>100	unlikely

10. REFERENCES

10.1. Mehlich, A. 1972. Uniformity of expressing soil test results on a volume basis. Comm. Soil Sci. Plant Analysis. 3:417-424.

10.2. Mehlich, A. 1984. Mehlich 3 soil test extractant: A modification of Mehlich 2 extractant. Comm. Soil Sci. Plant Anal. 15(12): 1409-1416.

10.3. Murphy, J., and J. R. Riley. 1962. A modified single solution method for the determination of phosphate in natural waters. Anal. Chem. Acta. 27:31-36.

10.4. Watanabe, F. S., and S. R. Olsen. 1965. Test of an ascorbic acid method for determining phosphorus in water and $NaHCO_3$ extracts from soil. Soil Sc. Soc. Am. Proc. 29:677-678.

10.5   Mehlich, A. 1978. New extractant for soil test evaluation of phosphorus, potassium, magnesium, calcium, sodium, manganese and zinc.   Comm. Soil Sci. Plant Anal. 9(6):477-492.

DETERMINATION OF CALCIUM, POTASSIUM, MAGNESIUM AND SODIUM BY MEHLICH NO. 3 EXTRACTION

1. PRINCIPLE OF THE METHOD

1.1 The extraction of calcium, magnesium, potassium and sodium by this method is designed to be applicable across a wide range of soil properties ranging in reaction from acid to basic. The Mehlich No. 3 method correlates well with Melich No. 1 Mehlich No. 2 and ammonium acetate (see 9.1). For specific extraction values and correlation coefficients, see Mehlich (9.1 and 9.2).

1.2 This extractant was developed by Dr. Adolf Mehlich in 1980-81, and was described by Hatfield for Mehlich (see 9.2). This procedure was developed on a 1:10 soil-solution ratio (2.5 cm^3 soil + 25 cm^3 extractant) for a 5-minute shaking period at 200, 4-cm reciprocations/minute.

2. RANGE AND SENSITIVITY

2.1 Following a 1:4 dilution of the soil extract with the Lithium Working Solution (see 6.5), soil concentrations of potassium and sodium can be determined up to 1564 and 920 kg/ha, respectively. Following a 1:10 dilution of the soil extract with the Lanthanum Solution (see 6.6), soil concentrations of calcium and magnesium can be determined up to 10,020 and 1216 kg/ha, respectively. An atomic absorption spectrophotometer equipped with a 3-standard microprocessor is required to obtain linearity at instrument readings above 100. In the absence of microprocessor-equipped instrumentation, extractable calcium and magnesium can be determined in soil concentrations up to 10 and 2 meq/100 cm^3, respectively.

2.2 Sensitivity will vary with the type of instrument, wavelength selection and method of excitation.

2.3 The commonly used methods of analysis for the above elements are flame emission and atomic absorption spectrophotometer. For a complete description of these methods, see Isaac and Kerber (9.3).

3. INTERFERENCES

3.1 Chemical interferences and compensation for changes in the characteristics of the extract to be analyzed must be acknowledged. Internal standards such as lithium and compensating elements such as lanthanum are required for most flame methods of excitation (see Isaac and Kerber, 9.3).

4. PRECISION AND ACCURACY

4.1   Repeated analyses of the same soil with medium ranges of potassium, calcium and magnesium will give variances from 5% to 10%. The major portion of the variance is related more to the heterogeneity of the soil than to measurement, extraction, or method of analysis.

5. APPARATUS

5.1   No. 10 (2-mm) sieve.

5.2   2.5 cm^3 scoop (volumetric) and teflon-coated leveling rod.

5.3   100-mL extraction bottles, plastic or glass.

5.4   Reciprocating shaker (200, 4-cm reciprocations/minute).

5.5   Filter funnels, 11 cm.

5.6   Whatman No. 1 (or equivalent) filter paper, 11 cm.

5.7   Funnel rack.

5.8   Vials, polystyrene plastic, 25-mL capacity, for collection of extract and sample dilutions

5.9   Automatic dispenser for extractant, 25-mL capacity.

5.10   Diluter-dispenser mechanism or pipettes, 10-mL capacity.

5.11   Flame emission, atomic absorption and/or ICP spectrophotometer. A 3-standard microprocessor for the atomic absorption spectrophotometer is desirable.

5.12   Volumetric flasks and pipettes as required for preparation of reagents and standard solutions.

5.13   Analytical balance.

6. REAGENTS

6.1   All reagents are ACS analytical grade or standard reference solutions unless otherwise stated.

6.2   Extracting Reagent (0.2N CH_3COOH; 0.25N H_4NO_3; 0.015N NH_4F; 0.013N HNO_3; 0.001M EDTA - See 6.3 and 6.4 for the mixing procedure of the Extracting Reagent.

6.3   Ammonium Fluoride - EDTA Stock Reagent - Add approximately 600 mL pure water to a 1-liter volumetric flask. Add 138.9 g ammonium fluoride (NH_4F) and dissolve, then add 73.05 g EDTA, dissolve mixture and bring to volume with pure water.

6.4   Final Extraction Reagent Mixture - Add approximately 3000 mL pure water to a 4-liter volumetric flask, add 80 g ammonium nitrate (NH_4NO_3) and dissolve. Then add 16 mL NH_4F-EDTA stock reagent (see 6.3) and mix well. Add 46 mL glacial acid (CH_3COOH) and 3.28 mL conc nitric acid (HNO_3), then bring to volume with pure water and mix thoroughly. The final pH should be 2.5 ± 0.1.

6.5   Lithium Working Solution - Dilute 12.5 mL commercial lithium standard (1500 meq Li/L) to 1 liter with pure water. This solution is used as an internal standard for the determination of potassium.

6.6   Lanthanum Compensating Solution - Suspend 13 g lanthanum oxide (La_2O_3) in 50 mL pure water in a large beaker and dissolve with 28 mL conc nitric acid (HNO_3). Allow the solution to cool, and then pour it into a 2-liter volumetric flask and bring to volume with pure water.

6.7   Potassium and Sodium Standard - Dissolve 0.7456 g potassium chloride (KCl) and 0.5844 g sodium chloride (NaCl) in a liter volumetric flask and bring to volume with extractant (see 6.2). Alternatively, dilute 100 mL commercial potassium and sodium standard (100 meq K, 100 meq Na) to 1 liter with extractant (see 6.2). Prepare working standards to contain 0, 0.5, 1.0 and 2.0 meq K and Na/liter by appropriate dilution with extractant (see 6.2).

6.8   Calcium and Magnesium Standard - Dissolve 2.5 g calcium carbonate ($CaCO_3$) and 10.14 g magnesium sulfate ($MgSO_4 \cdot 7H_2O$) in approximately 500 mL extractant (see 6.2) and 10 mL conc nitric acid (HNO_3) and bring to 1 liter volume with extractant (see 6.2). Alternatively, dilute 500 mL commercial calcium reference standard (1 mL = 1 mg Ca) and 60.75 mL commercial magnesium reference standard (1 mL = 1 mg Mg) to 1 liter with extractant (see 6.2). Dilute with extractant (see 6.2) to obtain 5, 10, 15, 20 and 25 meq Ca, and 1, 2, 3 and 5 meq Mg/liter, respectively. Before making the 5, 10 and 15 meq standards to volume, add 3.0 mL conc nitric acid (HNO_3) to prevent calcium precipitation. The latter dilutions compose the working standards.

7. PROCEDURE

7.1 <u>Extraction</u> - Measure 2.5 cm^3 air-dry, 10 mesh (2-mm) soil into a 100-mL extraction bottle (see 5.3). Add 25 cm^3 of Extracting Solution (see 6.2) and shake for 5 minutes on a reciprocating shaker (see 5.4). Filter and collect the extract in 25 cm^3 plastic vials (see 5.8).

7.2 <u>Determination of Potassium and Sodium</u> - Using a diluter-dispenser (see 5.10) or pipette, transfer 2 mL soil extract (see 7.1) or Potassium-Sodium Working Standards (see 6.7) and 8 mL Lithium Working Solution (see 6.5) into plastic vials (see 5.8). Set the instrument reading at 100, using the 1-meq K-Na working standard. Atomize soil extract and record the instrument reading. For final calculations for a 20-cm depth of soil to meq/100 cm^3, mg/dm^3, kg/ha and lbs/acre of potassium and sodium, see 7.4.

7.3 <u>Instrument Calibration</u> - Proper precautions should be taken to follow the manufacturer's recommendations in the operation and calibration of the instrument. Linearity between the concentration of potassium and sodium can be ascertained by running a series of appropriate standards (see 6.7). If the instrument reading exceeds 200, dilute equal portions of the soil extract-lithium sample mixture with zero standard (see 7.2) and multiply the results by 2.

7.4 <u>Calculations (Potassium and Sodium)</u> - With the 1 meq Potassium-Sodium Standard (see 6.7), set the instrument at 100. Multiply the instrument reading (IR) by 0.01 meq K or Na/100 cm^3 soil. Alternatively, the IR x 3.91 = mg K/dm^3 and IR x 2.3 = mg Na/dm^3 of soil. To convert mgK and Na/dm^3 to kg/ha, multiply by 2. Multiply mgK or Na/dm^3 by 1.78 to obtain lbs/acre to a depth of 20 cm. All of the above calculations are based on a volume of soil (see 5.2) which is employed in these soil test procedures (see 9.5).

7.5 <u>Determination of Calcium and Magnesium</u> - Using a diluter-dispenser (see 5.10) or pipette, transfer 1 mL Calcium-Magnesium Standard (see 6.8) or soil extract (see 7.1) and 9 mL Lanthanum Compensating Solution (see 6.6) into plastic vials (see 5.8). Adjust the instrument to zero with a blank composed of 1 mL extractant (see 6.4) and 9 mL of Lanthanum Compensating Solution (see 6.6). Standardize the instrument with Ca-Mg Standards (see 6.8) by setting the 10 meq Ca-2 meq Mg Working Standard (see 6.8) at 100 and the 25 meq Ca - 5 meq Mg Working Standard at 250 instrument readings, respectively. An instrument equipped with a 3-standard microprocessor is required in order to obtain linearity above an instrument reading of 100.

7.6 Instrument Calibration - Proper precautions should be taken to follow the manufacturer's recommendations in the operation and calibration of the instrument. Linearity between the concentration of calcium and magnesium can be ascertained by running a series of appropriate standards (see 6.8). Calcium and magnesium concentrations are linear up to 10 meq Ca and 2 meq Mg at a corresponding instrument reading of 100. By the use of a 3-standard microprocessor, linearity can be obtained up to 25 meq Ca and 5 meq Mg with a corresponding instrument reading of 250. If scale expansion above the 100 instrument reading is not available and the unknown readings exceed 100, dilute the known aliquot of soil extract-lanthanum mixture with zero standard (see 6.4) and multiply by the dilution factor.

7.7 Calculations (Calcium and Magnesium) - With the 10 meq Ca- 2 meq Mg and 25 meq Ca- 5 meq Mg standards (see 6.8), set at an instrument reading (IR) of 100 and 200 respectively. The corresponding concentrations on a volume soil basis are: IR x 0.1 = meq Ca/100 cm^3 and IR x 0.02 = meq Mg/100 cm^3. Alternatively, to convert IR to mg Ca and Mg/dm^3, multiply by 20.04 and 2.432, respectively. To obtain kg Ca and Mg/ha, multiply mg/dm^3 by 2. Then kg/ha of Ca and Mg multiplied by 0.891 = lbs/acre.

8. INTERPRETATION

8.1 Evaluation of the analytical results and the corresponding fertilizer recommendations must be based on field response data conducted under local soil-climate crop conditions (see 9.6). Mehlich proposed critical levels of phosphorus, potassium, magnesium, manganese, zinc and copper for the Mehlich No. 3 extractant (see 9.2) as well as interpretative guidelines for evaluating percentage of calcium and base saturation (see 9.1).

9. REFERENCES

9.1. Mehlich, A. 1978. New extractant for soil test evaluation of phosphorus, potassium, magnesium, calcium, sodium, manganese and zinc. Comm. Soil Sci. Plant Anal. 9: 477-492.

9.2. Mehlich, A. 1984. Mehlich 3 soil test extractant: A modification of Mehlich 2 extractant. Comm. Soil Sci. Plant Anal. 15: 1409-1416.

9.3   Isaac, R. A., and J. D. Kerber. 1972. Atomic absorption and flame photometry: Techniques used in soil, plant, and water analysis, pp. 17-38. IN: L. M. Walsh (ed.), Instrumental Methods for Analysis of Soils and Plant Tissue. Soil Science Society of America. Madison, WI.

9.4   Mehlich, A. 1953. Determination of P, Ca, Mg, K, Na and NH_4 (mimeo). North Carolina Soil Test Division., Raleigh, NC.

9.5   Mehlich, A. 1972. Uniformity of expressing soil test results. A case for calculating results on a volume basis. Comm. Soil Sci. Plant Anal. 3:417-424.

9.6   Doll, E. C., and R. E. Lucas. 1973. Testing soils for potassium, calcium and magnesium, pp. 133-152. IN: L.M. Walsh and J. D. Beaton (eds.), Soil Testing and Plant Analysis. Soil Science Society of America., Madison, WI.

DETERMINATION OF MANGANESE, ZINC AND COPPER BY MEHLICH NO. 3 EXTRACTION

1. PRINCIPLE OF THE METHOD

1.1 The extraction and determination of manganese, zinc and copper by this procedure is applicable across a wide range of soil properties ranging in reaction from acid to basic. Although the method was correlated with established extractants from several regions and critical levels were established, the specific critical levels should be based on local soil, crop and climatic conditions. Good correlations were obtained between Mehlich No. 2 and Mehlich No. 3 for manganese and zinc (see 9.2), even though the mean values were not the same. Mehlich No. 3 correlated well with the Mehlich-Bowling extractant for copper (see 9.5). For critical levels for the Mehlich No. 3 extractant, see 9.2.

2. RANGE AND SENSITIVITY

2.1 Manganese, zinc and copper can be extracted and determined without dilution in soil concentrations of 20, 4.0 and 4.0 mg/dm^3, respectively. This equates to 40, 8.0 and 8.0 kg/ha for manganese, zinc and copper, respectively. Higher concentrations can be determined with appropriate dilutions or by using instrumentation equipped with a 3-standard microprocessor. The method was developed using atomic absorption spectrophotometry at a 1:10 soil-to-solution ratio (see 9.2).

2.2 Sensitivity will vary with the type of instrument, wavelength selection and method of excitation. For a complete description of these parameters, see Isaac and Kerber (9.8).

3. PRECISION AND ACCURACY

3.1 Repeated analyses of an internal check sample from 20 extractions by the NCDA Soil Testing Laboratory gave variances of 9.69%, 10.82% and 9.44% for manganese, zinc and copper, respectively. The mean values were 5.42, 2.29 and 1.71 mg/dm^3 of manganese, zinc and copper. The variance is essentially a factor related to sample heterogeneity rather than to measurement, extraction, or method of analysis.

4. INTERFERENCES AND CONTAMINATION

4.1 There are no known interferences.

4.2   The possibility of contamination between samples or from external sources (extraction vials, filter funnels and washing apparatus) should be recognized. Precautions should be taken to avoid the use of extraction vials which contain micronutrient impurities. Certain plastic bottles are also charged and can retain copper and zinc from previous extractions. Consequently, all laboratory apparatus must be washed with a reagent capable of displacing absorbed micronutrients. The rinsing solution used in this procedure is described below (see 7.0).

5. APPARATUS

5.1   No. 10 (2-mm) stainless steel sieve.

5.2 $2.5 cm^3$ scoop (volumetric) and teflon-coated leveling rod.

5.3   100-mL extraction bottles, preferably plastic.

5.4   Reciprocating shaker (200, 4-cm reciprocations/minute).

5.5   Plastic filter funnels, 11 cm.

5.6   Whatman No. 1 (or equivalent) filter paper, 11 cm.

5.7   Funnel rack.

5.8   Vials, polystyrene plastic, 25-mL capacity, for sample collection.

5.9   Automatic dispenser for extractant, 25-mL capacity.

5.10   Atomic absorption spectrophotometer and/or ICP. A 3-standard microprocessor AA is desirable.

5.11   Volumetric flasks and pipettes as required for preparation of reagents and standard solution.

5.12   Analytical balance.

6. REAGENTS

6.1   All reagents are ACS analytical grade or standard reference solutions unless otherwise stated.

6.2   <u>Extracting Reagent</u> - (0.2N CH_3COOH; 0.25N NH_4NO_3; 0.015N NH_4F; 0.013N HNO_3; 0.001M EDTA). See 6.3 and 6.4 for the mixing procedure of extraction reagents.

6.3   Ammonium Fluoride - EDTA Stock Reagent - Add approximately 600 mL pure water to a 1-liter volumetric flask; add 138.9 g ammonium fluoride (NH_4F) and dissolve. Then add 73.05 g EDTA. Dissolve the mixture and bring it to volume with pure water.

6.4   Final Extraction Reagent Mixture - Add approximately 3000 mL pure water to a 4-liter volumetric flask, add 80 g ammonium nitrate (NH_4NO_3) and dissolve. Add 16 mL NH_4F-EDTA stock reagent (see 6.3) and mix well. Add 46 mL glacial acetic acid (CH_3COOH) and 3.28 mL conc nitric acid (HNO_3), then bring to volume with pure water and mix thoroughly. The final pH should be 2.5 $\pm$ 0.1.

6.5   Manganese Standard (20 mg Mn/L) - Dilute 20 mL commercial Manganese Reference Standard (1 mL = 1 mg Mn) to 1 liter with the extractant (see 6.4).

6.6   Manganese Working Standards - Dilute 25, 50, 75 and 100 mL Manganese Standard (see 6.5) to 1 liter with extractant (see 6.4), corresponding to 0.5, 1.0, 1.5 and 2.0 mg Mn/L. Following the manufacturer's guidelines, set the instrument at zero with the extractant (see 6.4). Using the 2.0 mg Mn/liter standard, set the instrument reading to 100. Intermediate standards can be used to check linearity. Higher concentration ranges can be used with an atomic absorption spectrophotometer equipped with a 3-standard microprocessor.

6.7   Procedure and Calculations - With the 2.0 mg Mn/L, set the instrument reading to 100. Soil extracts can be read directly with appropriate dilutions when instrument readings exceed 100. Instrument readout x 0.2 = mg Mn/dm^3 of soil. The mg Mn/dm^3 x 2 = kg Mn/ha and kg Mn/ha x 0.891 = lbs Mn/acre. These calculations are based on a volume of soil to a depth of 20 cm. For the rationale, see 9.4 and 9.7.

6.8   Zinc and Copper Standards (4 mg Zn, Cu/L) - Dilute 100 mL commercial reference standard (1 mL = 1 mg Zn, 1 mg Cu/mL) to 1 liter with extractant (see 6.4). From this mixture, dilute 40 mL to 1 liter with extractant (see 6.4) for corresponding concentrations of 4 mg Zn and Cu/L. These standards can be prepared separately or in combination with each other, depending on preference.

6.9 Zinc and Copper Working Standards - Dilute 5 and 10 mL of zinc and copper standard (4 mg Zn, Cu/L) to 1 liter with extractant corresponding to 0.2 and 0.4 mg Zn, Cu/L. Following the manufacturer's guidelines, adjust the instrument to zero with extractant (see 6.4). Atomize the 0.4 mg Zn, Cu/L standard and adjust the instrument reading to 100. Intermediate standards can be prepared to check for linearity. Higher concentrations can be used on instruments equipped with a 3-standard microprocessor.

6.10 Procedure and Calculations - With the 0.4 mg Zn, Cu/L standard, set the instrument reading to 100. Soil extracts can be read directly with appropriate dilutions when the instrument reading exceeds 100. Instrument reading x 0.4 = mg Zn, Cu/dm^3 of soil. Alternately, instrument reading x 0.008 = kg Zn, Cu/ha. To convert kg Zn, Cu/ha to lbs/acre, multiply by 0.891. These calculations are based on a volume of soil to a depth of 20 cm. For the rationale, see 9.4 and 9.7.

7. DECONTAMINATION SOLUTION (0.2% $AlCl_3 \cdot 6H_2O$)

7.1 Dissolve 20 g aluminium chloride ($AlCl_3$) in about 2 liters of water and make to 10 liters with pure water. This solution volume can vary with the number of samples involved.

7.2 Procedure - Wash the extraction bottles (see 5.3), extraction vials (see 5.8), and funnels (see 5.7) with hot tap water, rinse with 0.2% $AlCl_3 \cdot 6H_2O$, then rinse with distilled water. After placing filter paper into the funnels, rinse the paper with 0.2% $AlCl_3 \cdot 6H_2O$ followed with distilled water. Allow it to drain. All washing apparatus should be constructed from stainless steel or plastic.

8. INTERPRETATION

8.1 Manganese - Calibration of the manganese soil test with this extractant is based on extractable manganese and soil pH (see 9.1). Equations predicting the manganese availability index (MnAI) for soybeans and corn are as follows:

Soybean   MnAI = 101.2 + 0.6 (MnI) - 15.2 (pH)

Corn   MnAI = 108.2 + 0.6 (MnI) - 15.2 (pH)

The critical soil test MnI = 4 mg Mn/dm^3, which is equal to a 25 soil test index. Because of the limited soil test calibration for other crops, calculation of the MnAI for these crops is based on their sensitivity to manganese, as compared to corn or soybeans.

For example, the soybean MnAI is used to predict manganese needs for small grains, since their sensitivity is closely related to that of soybeans.

8.2   Copper - The critical copper soil test level was established with the Mehlich-Bowling (see 9.5) and the Mehlich No. 3 extractants (see 9.2). The critical level is 0.5 mg Cu/dm^3, which equates to a soil test index of 25.

8.3   Zinc - The critical zinc soil test level by this procedure is 1.0 mg Zn/dm^3, which equates to a soil test index of 25. A zinc availability index (ZnAI) has been established for mineral, mineral-organic, and organic soils and is based on the relationship between extractable zinc and soil pH (see 9.9). These values are as follows:

$$ZnAI \text{ (mineral soils)} = ZnI \times 1.0$$

$$ZnAI \text{ (mineral - organic soils)} = ZnI \times 1.25$$

$$ZnAI \text{ (organic soils)} = ZnI \times 1.66$$

9. REFERENCES

9.1.   Mascagni, H. J., and F. R. Cox. 1985. Calibration of a manganese availability index for soybeans soil test data. Soil Sci. Soc. Am. J. 49(2):382-386.

9.2.   Mehlich, A. 1984. Mehlich 3 soil test extractant: A modification of Mehlich 2 extractant. Comm. Soil Sci. Plant Anal. 15(12):1409-1416.

9.3.   Mehlich, A. 1978. New extractant for soil test evaluation of phosphorus, potassium, calcium, magnesium, sodium, manganese and zinc. Comm. Soil Sci. Plant Anal. 9(6):477-492.

9.4.   Mehlich, A. 1972. Uniformity of expressing soil test results. A case for calculating results on a volume basis. Comm. Soil Sci. Plant Anal. 3(5):417-424.

9.5   Mehlich, A., and S. S. Bowling. 1975. Advances in soil test methods for copper by atomic absorption spectrophotometry. Comm. Soil Sci. Plant Anal. 6(2):113-128.

9.6   Makarim, A. K., and F. R. Cox. 1983. Evaluation of the need for copper with several soil extractants. Agron. J. 75:493-496.

9.7   Mehlich, A. 1973. Uniformity of soil test results as influenced by volume weight. Comm. Soil Sci. Plant Anal. 4:475-486.

9.8   Isaac, R. A., and J. D. Kerber. 1972. Atomic absorption and flame photometry. Techniques and uses in soil, plant and water analysis, pp. 17-38. IN: L. M. Walsh (ed.), Instrumental Methods for Analysis of Soils and Plant Tissue. Soil Science Society of America, Madison, WI.

9.9   Junus, Mohd Aris. 1984. Incorporation of acidity and cation exchange capacity in the zinc soil test interpretation. Ph.D. dissertation, Soil Sci. Dept., North Carolina State University, Raleigh, NC.

DETERMINATION OF BORON BY HOT WATER EXTRACTION

1. PRINCIPLE OF THE METHOD

1.1 This method determines the amount of available soil boron. It was first proposed by Berger and Truog (see 12.1) and given in detail by Wear (see 12.4) and Bingham (see 12.10). Various modifications of the technique have been studied by Gupta (see 12.3) and Odom (see 12.8). Wolf (see 12.6, 12.7) has developed a different extraction method using Azomethine-H as the color development reagent. However, hot water extraction is a method in common use, although other methods have been proposed (see 12.10).

2. RANGE AND SENSITIVITY

2.1 Boron can be extracted and determined in soil concentrations from less than 1 ppm to 10 ppm without dilution. The upper limit may be extended by diluting the extract prior to colorimetric determination.

2.2 The technique for extraction involves boiling the soil in water for a specified time period. The effect of time and boiling technique has been studied by Gupta (see 12.3) and Odom (see 12.8). A mechanized technique has been developed by John (see 12.5).

3. INTERFERENCES

3.1 Obtaining a clear filtrate after boiling may be a problem. The use of double filter paper (see 7.1) is helpful. Some have proposed boiling soil in a 0.1% solution of calcium chloride ($CaCl_2 \cdot 2H_2O$) to help clear the filtrate (see 12.9).

3.2 Interferences may occur in the color development when various color reagents (see 12.2, 12.3, 12.4, 12.9) other than those proposed here are used. With the use of Azomethine-H, interferences are minimal (see 12.6, 12.7).

4. PRECISION AND ACCURACY

4.1 Variance is essentially a factor related to the heterogeneity of the soil rather than to the extraction or colorimetric procedures. However, care in mixing in the Azomethine-H procedure is important (see 12.7).

5. APPARATUS

5.1   No. 10 (2-mm) sieve.

5.2   8.5-cm^3 scoop, volumetric.

5.3   Refluxing apparatus, 250-mL capacity.

5.4   Hot plate.

5.5   Filter funnel, 11 cm.

5.6   Whatman No. 42 filter paper or equivalent, 11 cm.

5.7   Photoelectric colorimeter set at 430 nm.

5.8   Volumetric flasks and pipettes as required for preparation of reagents, standard solutions and color development.

5.   Analytical balance.

6. REAGENTS

6.1   <u>Azomethine-H Reagent</u> - Dissolve 0.9 g Azomethine-H and 2 g ascorbic acid in 10 mL pure water with gentle heating in a water bath. When it is dissolved, dilute 100 mL with pure water. If the solution is turbid, reheat in the water bath until it is clear. The reagent will last 14 days when it is refrigerated.

6.2   <u>Buffer Masking Reagent</u> - Dissolve 250 g ammonium acetate $(NH_4C_2H_3O_2)$, 25 g tetrasodium salt of (ethylenedinitrillo) tetraacetic acid and 10 g disodium salt of nitrilotriacetic acid in 400 mL pure water. Slowly add 125 mL of acetic acid $(HC_2H_3O_2)$.

6.3   <u>Boron Standard (100 ppm B)</u> - Weigh 0.5716 g boric acid (H_3BO_3) into a 1-liter volumetric flask and bring to volume with pure water.

7. PROCEDURE

7.1   <u>Extraction</u> - Weigh 10 g or scoop 8.5-cm^3 (see 5.2) of air-dry <10-mesh (2-mm) soil into a refluxing flask and add 20 mL pure water. Assemble the refluxing apparatus and place the flasks on the hot plate. Bring to a boil and boil 10 minutes. Filter through double filter paper (see 5.6) and collect the filtrate for boron determination.

7.2　Color Development - Pipette 4 mL extractant into a spectrophotometer cuvet. Add 1 mL Buffer Masking Reagent (see 6.2) and 1 mL Azomethine-H Reagent (see 6.1). Add Azomethine-H Reagent and mix the solution immediately. Cap and invert the cuvet and let stand for 1 hour. Read the transmittance (% T) at 430 nm with the blank being water.

8. CALIBRATION AND STANDARDS

8.1　Working Boron Standards - With the Boron Standard (see 6.3), prepare 5 working standards giving 0.125, 0.25, 0.5, 1.0 and 2.0 ppm boron. Use 4 mL aliquots of each standard and carry through the color development (see 7.2).

8.2　Calibration Curve - On semi-log graph paper, plot the percent transmittance on the logarithmic scale versus ppm boron on the linear scale.

9. CALCULATION

9.1　The results are reported as ppm B in soil: ppm in soil = ppm in the extractant x 2. If the extraction filtrate is diluted, the dilution factor must be applied.

10. EFFECTS OF STORAGE

10.1　Soils may be stored in an air-dry condition for several months with no effect on extractable boron.

10.2　After extraction, the extraction solution containing boron should not be stored any longer than 24 hours.

11. INTERPRETATION

11.1　Accurate fertilizer recommendations for boron must be based on known field responses based on local soil-climate-crop conditions (see 12.1, 12.2, 12.3, 12.4 12.9). For most soils and crops, the amount of boron extracted should be interpreted as follows:

Category	ppm B in soil
Insufficient	<1.0
For normal growth	1.0-2.0
High	2.1-5.0
Excessive	5.0

12. REFERENCES

12.1. Berger, K. C., and E. Truog. 1940. Boron deficiency as revealed by plant and soil tests. J. Am. Soc. Agron. 32:297-301.

12.2. Berger, K. C. and E. Truog. 1944. Boron tests and determination for soils and plants. Soil Sci. 57:25-36.

12.3. Gupta, U. C. 1967. A simplified method for determining hot water-soluble boron in podzol soils. Soil Sci. 103:424-428.

12.4. Wear, J. I. 1965. Boron, pp. 1059-1063. IN : C. A. Black (ed.), Methods of Soil Analysis, Part 2, Agron. No. 9. American Society of Agronomy, Madison, WI.

12.5 John, M. K. 1973. A batch-handling technique for hot-water extraction of boron from soils. Soil Sci. Soc. Amer. Proc. 37:332-333.

12.6 Wolf, B. 1971. The determination of boron in soil extracts, plant materials, composts, manures, water and nutrient solutions. Comm. Soil. Sci. Plant Anal. 2(5): 363-374.

12.7 Wolf, B. 1974. Improvements in the Azomethine-H method for the determination of boron. Comm. Soil Soc. Plant Anal. 5(1):39-44.

12.8 Odom, J. W. 1980. Kinetics of the hot-water soluble boron soil test. Comm. Soil Sci. Plant Anal. 11(7):759-769.

12.9 Reisenauer, H. M., L. M. Walsh, and R. G. Hoeft. 1973. Testing soils for sulfur, boron, molybdenum, and chlorine, pp. 173-200. IN: L. M. Walsh and J. D. Beaton (eds.), Soil Testing and Plant Analysis, rev. ed. Soil Science Society of America, Madison, WI.

12.10 Bingham, F. T. 1982. Boron, pp 431-447. IN: A. L. Page (ed.), Methods of Soil Analysis, Part 2, Chemical and Microbiological Properties, 2nd ed. American Society of Agronomy, Madison, WI.

DETERMINATION OF ZINC BY 0.1N HCl EXTRACTION

1. PRINCIPLE OF THE METHOD

1.1 This method is primary for determining extractable zinc in acid soils (pH less than 7.0). The test is designed to divide soils into two groups: those which cannot supply the crop requirement and therefore will require zinc and those which have an adequate supply of zinc to meet the crop requirement. The method is not suitable for alkaline soils unless additional measurements are made (see 12.1, 12.2 and 12.3).

1.2 This procedure as presented is a modification of a method used by Wear and Evans (see 12.4) with early calibration work by Gilroy (see 12.5). Many variations of the method have been used. The main differences between the methods include modifications in shaking time and soil:extractant ratios.

1.3 The procedure is based upon the assumption that all or a portion of the soil zinc which will become available for plant uptake during a growing season is acid soluble. The quantity of acid soluble zinc serves as an index of availability (see 12.6).

2. RANGE AND SENSITIVITY

2.1 The sensitivity of the analytical procedure used to determine the concentration of zinc in the extract will depend on the method of analysis (see 12.7 and 12.11). Normally, the zinc content of a soil can be determined at 0.5 mg Zn/kg or less.

3. INTERFERENCES

3.1 In the usual range of concentration of zinc and other elements in acid soils (pH less than 7.0), there are no interferences which would affect the determination. The acid extractant can be neutralized by free carbonates and rendered ineffective for alkaline soils.

3.2 All apparatus that will come in direct contact with the extractant, soil or extraction filtrate must be thoroughly washed and rinsed in dilute redistilled hydrochloric acid (HCl) and zinc-free water before use. Avoid contact with rubber and metals.

4. PRECISION AND ACCURACY

4.1 Repeated extractions of soils in the range of 2 to 10 mg Zn/kg have given coefficients of variation of less than 10%. As with most soil tests, the major source of variation lies with sampling rather than with the analytical technique.

5. APPARATUS

5.1   No. 10 (2-mm opening) stainless steel sieve.

5.2   4.25-cm^3 scoop, volumetric.

5.3   Extraction bottle or flask, 50-mL capacity.

5.4   Mechanical reciprocating shaker, 180 oscillations per minute.

5.5   Filter funnel, 5.5 cm.

5.6   Whatman No. 2 filter paper (or equivalent), 9 cm.

5.7   Beaker, polypropylene, 30-mL capacity.

5.8   Funnel racks.

5.9   Atomic absorption or ICP emission spectrophotometer.

5.10 Volumetric flasks and pipettes for reagent and standard preparation.

5.11 HCl redistillation apparatus.

5.12 Mixed bed demineralizer.

5.13 Analytical balance.

6. REAGENTS

6.1   Zinc-Free Demineralized Water

6.2   Redistilled 6N HCl - Redistill a mixture of 1:1 conc hydrochloric acid (HCl) and pure water.

6.3   Extraction Reagent - Dilute 16.7 mL redistilled 6N hydrochloric acid (HCl) to 1 liter. Titrate with standard base to the phenolphthalein end point. Dilute acid or pure water should be added to obtain a 0.1N solution.

6.4   Zinc Standard (1000 mg/L) - Dissolve 1000 g pure zinc metal in 5 to 10 mL concentration hydrochloric acid (HCl) and dilute to 1 liter with pure water or use a commercially available atomic absorption zinc reference standard. Prepare working standards by diluting aliquots of the stock solution with the Extraction Reagent (see 6.3) to cover the working range. Working standards of 0, 0.1, 0.5, 1.0 and 2.0 mg Zn/L will be adequate.

7. PROCEDURE

7.1   Extraction - Weigh 5 g or scoop 4.25-cm^3 of air-dry <10-mesh soil into a 50-mL extraction flask. Add 20 mL Extraction Reagent (see 6.3) and shake for 30 minutes on a reciprocating shaker (see 5.4). Filter and collect the filtrate. A blank should be carried through the entire procedure with each run.

7.2   Analysis - The zinc concentration in the filtrate is determined by either atomic absorption or ICP emission spectrometer. The instrument manufacturer's instructions should be followed.

8. CALIBRATION AND STANDARDS

8.1   Working Standards - Working standards are prepared as in 6.4. If the zinc concentration exceeds that of the highest standard, either dilute the unknown or prepare another series of standards to cover the sample range.

8.2   Calibration - Calibration procedures vary with the type of instrument. The manufacturer's recommendations for operation and calibration should be followed.

9. CALCULATION

9.1   The results are reported as mg/L of extractable zinc in the soil, multiplying mg Zn/L in the filtrate x 8. If the extraction filtrate is diluted, the dilution factor should be applied. To convert to other units for comparison, see Mehlich (see 12.8).

10. EFFECTS OF STORAGE

10.1   Storage of soil in the air-dry condition in a closed container for several weeks should not affect the extractable zinc.

10.2   The extraction filtrate should not be stored for more than a few hours on the day of extraction.

11. INTERPRETATION

11.1   An evaluation of the results, as well as accurate fertilizer recommendations, must be based upon field response data conducted under local soil-climate-crop conditions. (see 12.9).

11.2   This procedure is a routine soil test used in Missouri. Field studies have shown that soils with less than 2 mg 0.1N HCl extractable Zn/kg will probably need zinc soil fertilization to obtain optimum zinc levels for corn and grain sorghum. Cox and Wear (see 12.10) found 3.1 kg/ha Zn or better in the soil adequate for corn on sandy soils.

12. REFERENCES

12.1. Nebraska Soil Testing Service. 1967. HCl extractable soil test for zinc (mimeo). University of Nebraska, Lincoln, NE.

12.2. Nelson, J. L., L. C. Boawn, and F. G. Viets, Jr. 1959. A method for assessing zinc status of soils using acid extractable zinc and "titratable alkalinity" values. Soil Sci. 88:275-283.

12.3. Viets, Jr., F. G., and L. C. Boawn. 1965. Zinc. IN: C. A. Black (ed.), Methods of Soil Analysis, Part 2. Agronomy No. 9. American Society of Agronomy, Madison, WI.

12.4. Wear, J. I., and C. E. Evans. 1968. Relationship of zinc uptake by corn and sorghum to soil zinc measured by three extractants. Soil Sci. Soc. of Am. Proc. 32:543-546.

12.5 Gilroy, T. E. 1969. Relationship between extractable soil zinc and zinc concentration in corn leaves. M. S. thesis, University of Missouri, Columbia, MO.

12.6 Tucker, T. C., and L. T. Kurtz. 1955. A comparison of several chemical methods with the bio-assay procedure for extracting zinc from soils. Soil Sci. Soc. Am. Proc. 19:477-481.

12.7 Isaac, R. A., and J. D. Kerber. 1972. Atomic absorption and flame photometry: Techniques and uses in soil, plant and water analysis, pp. 17-38. IN: L. M. Walsh (ed.), Instrumental Methods of Analysis of Soils and Plant Tissue. Soil Science Society of America, Madison, WI.

12.8 Mehlich, A. 1972. Uniformity of expressing soil test results: A case for calculating results on a volume basis. Comm. Soil Sci. Plant Anal. 3:417-424.

12.9 Viets, Jr., F. G., and W. L. Lindsay. 1973. Testing soils for zinc, copper, manganese and iron, pp. 153-172. IN: L. M. Walsh and J. D. Beaton (eds.), Soil Testing and Plant Analysis, rev. ed. Soil Science Society of America, Madison, WI.

12.10 Cox, F. R., and J. I. Wear (eds). 1977. Diagnosis and correlation of zinc problems in corn and rice production. Southern Cooperative Series Bull. 222, North Carolina State University, Raleigh, NC.

12.11 Soltanpour, P. N., J. B. Jones, Jr., and S. M. Workman. 1982. Optical emission spectrometry, pp. 29-65. <u>IN:</u> A. L. Page et al. (eds.) Methods of Soil Analysis, Part 2, rev. ed. American Society of Agronomy, Madison, WI.

DETERMINATION OF ACID-EXTRACTABLE COPPER BY THE MEHLICH-BOWLING METHOD

1. PRINCIPLE OF THE METHOD

1.1   Determination of copper by this method involves extraction with 0.5N HCl plus 0.05 $AlCl_3$ using a 1:5 soil: extractant ratio by volume, a shaking period of 5 minutes (see10.1), and subsequent determination by atomic absorption spectrophotometry. For meaningful analysis, concentrations of 0.01 µg Cu/mL were required for direct aspiration. With instrumentation short of this requirement, a single complexing-organic reagent consisting of 2 g APDC (ammonium-1-pyrrolidine dithiocarbamate) dissolved in 600 mL 95% ethanol and then mixed with 400 mL n-butyl acetate has been developed. This reagent renders the method expeditious, increases storage stability for APDC up to 6 months, and provides improved separation of the organic phase and extractant.

1.2   The method has been tested on soils from numerous field experiments and correlated with crop response using wheat and soybeans, and for sorghum on organic soils (Histosols) and wheat on mineral soils (Ultisols). The probability of crop response to Cu based on extractable Cu level has been determined for use in North Carolina.

2. RANGE AND SENSITIVITY

2.1   The procedure involving use of organic phase separation depends on the sensitivity of the atomic absorption spectrophotometer for Cu. For a meaningful detection and range of Cu with the proposed method, minimum sensitivity was found to be 0.01 µg Cu/mL with linearity up to 1.0 µg Cu/mL. If the instrument does not meet this requisite, aspiration from the organic phase as proposed in the procedure readily meets these requirements.

3. INTERFERENCES

3.1   Interferences relate primarily to "colloidal" substances extracted from certain Histosols, which tend to form a gel in the organic phase. However, the inclusion of ethanol in the procedure largely eliminates this possible interference. These unknown substances remain in solution and do not interfere when aspirated from the original extractant phase.

4. PRECISION AND ACCURACY

4.1   The major source of variance in the extraction is the hetero-geneity of the soil sample itself. Repeated analysis of the same sample normally gives a coefficient of variation of about 12%.

5. APPARATUS

5.1   Soil crusher, sieve (2-mm opening), stainless steel and all contact portion Cu free.

5.2   5-cm^3 scoop, volumetric (plastic or stainless steel).

5.3   Extraction bottle or flask, 60 mL wide mouth (4.5 x 8 cm).

5.4   Mechanical reciprocating shaker, 180 oscillations per minute or greater.

5.5   Wrist action type shaker.

5.6   Filter funnel, 11 cm.

5.7   Acid wash filter paper, fine porosity Whatman No. 42, 11 cm or equivalent.

5.8   Atomic absorption spectrophotometer with supporting materials, including fuel, oxidant and Cu hollow cathode lamp.

5.9   Funnel racks.

5.10   Analytical balance.

5.11   Volumetric flasks, 125-mL Erlenmeyer flasks and pipettes as required for preparation of reagents and standard solutions.

5.12   Pyrex culture tubes, 20 mL (16 x 150 mm) with pressure-fitted inert linear screw caps.

6. REAGENTS

6.1   Extracting Reagent (0.5N HCl - 0.05N AlCl$_3$) - Dissolve 4.0 g AlCl$_3 \cdot$ 6H$_2$O in approximately 500 mL pure water, add 42 mL conc HCl. Dilute to volume (1 liter) with with pure water. Mix.

6.2   Copper Standard, 10 μg Cu/mL - Dilute 10 mL copper reference solution (1000 mg/mL) to 1 liter with the Extracting Reagent (see 6.1). Prepare standards to contain 7.5, 5.0, 2.5, 1.0 and 0 mg Cu/mL by appropriate dilutions with the Extraction Reagent (see 6.1).

6.3   Copper Standard, 1 µg Cu/mL - Dilute 10 mL Copper Standard (see 6.2) to 1 liter with the Extracting Reagent (see 6.1). Prepare appropriate standards by dilution with the Extracting Reagent (see 6.1) as needed.

6.4   APDC - Butyl Acetate - Ethanol Reagent - Dissolve 2 g ammonium-1-pyrrolidine dithiocarbamate in 600 mL of 95% ethanol. Add 400 mL n-butyl acetate (technical). Mix and store in a brown or amber bottle.

6.5   Working Standards for 0 and 1.0 mg Cu/mL (Organic Phase Aspiration) - Measure 50 mL Extracting Reagent (see 6.1) or 50 mL Copper Standard (see 6.3) into an 125-mL Erlenmeyer flask. Add 25 mL APDC Reagent (see 6.4). Seal with a stopper. Vigorously mix by hand for 15 minutes and then shake for 10 minutes on a wrist-action type shaker. Decant the organic phase into 25-mL glass cylinders, allow it to separate, and aspirate from the organic phase.

7. PROCEDURE

7.1   Extraction - Soil samples are dried at about $10°$ C above the ambient temperature, ground in a stainless steel mill and passed through a 2-mm stainless steel screen. Measure by volume (see 5.2) 5 cm^3 of soil into a 60-mL wide mouth bottle, add 25 mL Extracting Reagent (6.1) and shake for 5 minutes on a reciprocating shaker at a minimum of 180 oscillations per minute. Filter through an 11-cm fine porosity quantitative filter paper.

7.2   Direct Aspiration from Extraction in the 0.1 - 10.0 µg Cu/mL Range - If the instrument reading exceeds the maximum of 1 µg cu/mL, use the Copper Standard (see 6.2) for standardization and aspirate from the original soil extract. If the concentration of Cu is still too high, dilute the sample with the Extracting Reagent (see 6.1).

7.3   Direct Aspiration from Extractant in the 0.01 - 1.0 µg Cu/mL Range - If the instrument has the detection limit corresponding to 0.01 µg/mL, use the Copper Standard (see 6.3) for standardization and aspirate from the original soil extract.

7.4   Aspiration from Organic Phase - Pipette 10 mL of soil extract into a 20 mL (16 x 150 mm) Pyrex culture tube with pressure-fitted inert liner screw caps. By means of a calibrated automatic glass pipette, add 5 mL APDC (see 6.4). Seal with a stopper and, within 5-6 minutes, mix vigorously by hand for 15 seconds, followed by 10 minutes of agitation on a wrist-action type shaker. Allow to separate and aspirate from the organic phase. Prepare the Working Standards as described in section 6.5.

8. CALIBRATION AND STANDARDS

8.1   Multiply appropriate solution concentrations (1:5 soil: extractant ratio) by 5 to obtain μg Cu/cm^3 of soil. To convert to kg/ha to a soil depth of 20 cm, multiply μg/cm^3 by 2. To convert to lb/A, multiply μg/cm^3 by 1.78.

9. CALCULATION

9.1	Response Probability	Extractable Cu μg/cm^3
	Highly probable	<0.5
	Probable	0.5 - 0.8
	Probable on sensitive crops	0.9 - 1.2
	Improbable	>1.2

10. REFERENCE

10.1. Mehlich, A., and S. S. Bowling. 1975. Advances in soil test methods for copper by atomic absorption spectrophotometry. Comm. Soil Sci. Plant Anal. 6:113-128.

AMMONIUM BICARBONATE-DTPA SOIL TEST FOR POTASSIUM, PHOSPHORUS, ZINC, IRON, MANGANESE, COPPER, AND NITRATE

1. PRINCIPLE OF THE METHOD

1.1   The Extraction Reagent is 1M ammonium bicarbonate (NH_4HCO_3) in 0.005M DTPA adjusted to a pH of 7.6 (see 12.1). Ammonium will exchange with potassium, bringing the latter into solution. The original pH of 7.6 allows DTPA to extract and chelate iron and other metals. Upon shaking, the pH rises due to the evolution of carbon dioxide. As the pH rises, the bicarbonate changes to carbonate. The carbonate ions precipitate calcium from calcium phosphate, thus increasing potassium solubility.

1.2   This method is highly correlated with the ammonium acetate method for potassium, the sodium bicarbonate method for phosphorus, and the DTPA method for zinc, iron, manganese, and copper (see 12.1).

2. RANGE AND SENSITIVITY

2.1   The range and sensitivity are the same as those for the DTPA, sodium bicarbonate, and ammonium acetate tests for micronutrients, phosphorus, and potassium, respectively.

3. INTERFERENCES

3.1   The Extracting Reagent is unstable with regard to pH and should be kept under mineral oil to prevent a pH change.

4. PRECISION AND ACCURACY

4.1   A coefficient of variability ranging from 5% to 10% can be expected for different determinations.

5. APPARATUS

5.1   No. 10 (2-mm opening) sieve constructed from stainless steel or nalgene.

5.2   Analytical balance.

5.3   125-mL polyethylene conical flasks.

5.4   Eberbach reciprocating shaker (or equivalent), 180 oscillations per minute.

5.5   Whatman No. 42 filter paper (or equivalent), 11 cm.

5.6   2.5-cm matching spectrometric tubes.

5.7   Atomic absorption spectrophotometer.

5.8   Photoelectric colorimeter suitable for measurement at 880 and 420 nm.

5.9   Funnel racks.

5.10   Accurate automatic diluter.

5.11   Volumetric flasks and pipettes as required for preparation of reagents, standard solutions and color development.

6. REAGENTS

6.1   Extracting Reagent - A 0.005M DTPA (diethylenetriamine-pentaacetic acid) solution is obtained by adding 1.97 g DTPA to 800 mL water. Approximately 2 mL of 1:1 NH_4OH is added to facilitate dissolution and to prevent effervescence when bicarbonate is added. When most of the DTPA is dissolved, 79.06 g NH_4HCO_3 are added and the solution is stirred gently until the added reagents are dissolved. The pH is adjusted to 7.6 with ammonium hydroxide (NH_4OH). The solution is diluted to 1.0 liter with pure water and is either used immediately or stored under mineral oil.

6.2   Mixed Reagent for Phosphorus - Dissolve 12.7 g ammonium molybdate $[(NH_4)_6 Mo_7 \cdot 4H_2O]$ in 250 mL pure water. Dissolve 0.2908 g antimony potassium tartrate $[K(SbO)C_4H_4O_6 \cdot 1/2 H_2O]$ in 1000 mL of 5N sulfuric acid (148 mL conc H_2SO_4 per liter). Mix the two solutions together thoroughly and make to a 2000-mL volume with pure water. Store in a pyrex bottle in a dark cool place.

6.3   Color Developing Reagent for Phosphorus - Add 0.739 g ascorbic acid to 140 mL of the Mixed Reagent. (This amount of reagent is enough for 24 phosphorus determinations, allowing 20 mL for wastage.) This reagent should be prepared as required, as it does not keep for more than 24 hours.

6.4   Antimony Sulfate Solution - 0.5 g antimony metal (Sb) is dissolved in 80 mL concentrated sulfuric acid (H_2SO_4), and 20 mL pure water is added to the acid carefully to prevent splattering. Heating will faciliate the dissolution of antimony metal in H_2SO_4. This solution is used for masking chlorides in nitrate determination.

6.5   Chromotropic Acid Solution (CTA) - A 0.00137M solution of CTA or 4,5-dihydroxy-2,7-naphthalene-disulfonic acid, disodium salt $[(HO)_2 C_{10}H_4 (SO_3Na)_2]$ is made by dissolving 0.5 g CTA in 4.0 kg concentrated sulfuric acid (H_2SO_4). This solution is used to develop color with nitrates.

6.6   Fisher G carbon black.

7. PROCEDURE

7.1   Extraction Method - Place 10 g soil (2-mm soil) in a 125-mL conical flask. Two 2.5 mL scoops Fisher G carbon black are added to each soil followed by 20 mL Extracting Reagent (see 6.1). The soil mixture is then shaken on an Eberbach reciprocal shaker for 15 minutes at 180 cycles/minute with flasks kept open. The extracts are then filtered through Whatman 42 filter paper (see 12.1 and 12.2).

7.2   Nitrates - Dilute a 0.3 mL aliquot of the soil extract to 3.5 mL with pure water using an automatic diluter. Add 2.0 mL antimony sulfate solution (see 6.4), followed by 6.5 mL CTA (see 6.5). Good mixing is necessary for consistent results. After 2 hours of cooling in water, the color intensity is read at 420 nm on a spectrophotometer (see 12.3 and 12.4).

7.3   Phosphorus - Dilute a 1.0 mL aliquot of the soil extract to 5.0 mL with pure water. Add 5 mL color developing reagent (see 6.3) carefully to prevent loss of sample due to excessive foaming. Add 15 mL pure water and stir. Let it stand for 15 minutes and measure the color intensity at 880 nm on a spectro-photometer (see 12.5). Standards are developed in one mL of extract in exactly the same way as described above.

7.4   Potassium - Potassium in the soil extract is determined directly by atomic absorption spectrometry using a potassium hollow cathode lamp. The 404 nm wavelength is used to reduce sensitivity. Standard solutions are made in Extracting Reagent (see 6.1).

7.5   Micronutrients - Zinc, iron, copper, and manganese are determined by atomic absorption spectrometry. The standard solutions of these metals are made in Extracting Reagent (see 6.1).

8. **STANDARDS**

8.1   Standards for potassium, zinc, iron, manganese, and copper are developed in the Extracting Reagent (see 6.1)

8.2   For the NO_3 and potassium standards, add 0.3 and 1 mL of Extracting Reagent and bring the volumes to 12 and 25 mL, respectively.

9. **CALCULATION**

9.1 NO_3-N, mg/kg in soil = NO_3-N, mg/kg in extract x 80.

P, mg/kg in soil = P, mg/kg in extract x 50.

K, mg/kg in soil = K, mg/kg in extract x 2.

Zn, Fe, Mn, and Cu, mg/kg in soil = Zn, Fe, Mn, Cu, mg/kg in extract x 2.

10. **EFFECTS OF STORAGE**

10.1   Air-drying and storage will not have any significant effect on the levels of nutrients.

10.2   The Extracting Reagent can be stored for 2 weeks under mineral oil. The pH can then be adjusted if necessary.

11. **INTERPRETATION**

11.1   The following tables give the index values and their interpretation for zinc, iron, copper, manganese, phosphorus, and potassium for the ammonium bicarbonate-DTPA soil test.

Index Values for Zn, Fe, Cu and Mn

Category	Zn	Fe	Cu	Mn
		mg/kg in soil		
Low	0.0-0.9	0.0-2.0	0.5	1.8
Marginal	1.0-1.5	2.1-4.0	---	---
Adequate	>1.5	>4.0	>0.5	>1.8

Index Values for P and K

Category	Sugar beets, corn, sorghum, small grains and grasses	Alfalfa
Very low	--	0- 3
Low	0- 3	4- 7
Medium	4- 7	8-11
High	8-11	12-15
Very high	>11	>15

mg/kg K in soil

Low	0- 60
Medium	61-120
High	>120

12. REFERENCES

12.1. Soltanpour, P. N., and A. P. Schwab. 1977. A new soil test for simultaneous extraction of macro- and micronutrients in alkaline soils. Comm. Soil Sci. Plant Anal. 8:195-207.

12.2. Soltanpour, P. N., A. Khan, and W. L. Lindsay. 1976. Factors affecting DTPA-extractable Zn, Fe, Mn and Cu from soils. Comm. Soil Sci. Plant Anal. 7:797-821.

12.3. West, P. W., and T. P. Ramachandran. 1966. Spectrophotometric determination of nitrate using chromotropic acid. Anal. Chem. Acta 350:317-324.

12.4. Sims, J. R., and G. D. Jackson. 1971. Rapid analysis of soil nitrate with chromotropic acid. Soil Sc. Soc. Am. Proc. 35: 603-606.

12.5 Watanabe, F. S., and S. R. Olsen. 1965. Test of an ascorbic acid method for determining phosphorus in water and $NaHCO_3$ extracts from soil. Soil Sc. Soc. Am. Proc. 29:677-678.

DETERMINATION OF ZINC, MANGANESE, IRON AND COPPER BY DTPA EXTRACTION

1. PRINCIPLE OF THE METHOD

1.1 The theoretical basis for the DTPA extraction is the equilibrium of the metal in the soil with the chelating agent. A pH level of 7.3 enables DTPA to extract Fe and other metals.

1.2 The use of DTPA as an extraction reagent was developed by Lindsay and Norvell (see 12.1).

2. RANGE AND SENSITIVITY

2.1 Zinc, iron, manganese, and copper can be extracted and determined in soil concentrations of 0.1 to 10 mg Zn/L, 0.1 to 10 mg Fe/L, 0.1 to 10 mg Mn/L, and 0.1 to 10 mg Cu/L without dilution. The range and upper limits may be extended by diluting the extracting filtrate prior to analysis.

2.2 The sensitivity will vary with the type of instrument used and wavelength selected.

2.3 The determination of copper, iron, manganese, and zinc in the filtrate is most commonly done by either atomic absorption or ICP emission spectrometry.

3. INTERFERENCES

3.1 Triethanolamine (TEA) is used to keep the pH close to 7.3.

3.2 All apparatus that will come in direct contact with the extractant and extraction filtrate of the soil must be thoroughly washed and rinsed in pure hydrochloric acid (HCl) and pure water before use. Avoid contact with rubber and metals.

3.3 Contamination of soil samples may occur in either the sampling equipment or soil-grinding equipment, especially for zinc and iron.

4. PRECISION AND ACCURACY

4.1 Repeated analysis of the same soil with medium concentration ranges of copper, iron, manganese, and zinc will give co-efficients of variability of from 10% to 15%. A major portion of the variance is related to the heterogeneity of the soil rather than to the extraction process or method of analysis.

5. APPARATUS

5.1 No. 10 (2-mm opening) sieve.

5.2 8.5-cm^3 scoop, volumetric.

5.3 Extraction flask, 125-mL polyethylene conical flasks.

5.4 Mechanical reciprocating shaker, 180 oscillations per minute.

5.5 Filter funnel, 11 cm.

5.6 Whatman No. 42 ashless filter paper (or equivalent), 11 cm.

5.7 Atomic absorption spectrophotometer.

5.8 pH meter, line or battery operated, with reproducibility to at least 0.1 pH unit and glass electrode pair with a calomel reference electrode.

5.9 Analytical balance.

5.10 Volumetric flasks, pipettes and microburet as required for preparation of reagents and standard solution.

6. REAGENTS

6.1 Extraction Reagent (DTPA -diethylenetriaminepentaacetic acid) - weigh 1.96 DPTA $\{[(HOCOCH_2)_2\ NCH_2\ CH_2]_2\ NCH_2$ COOH} into a 1-liter volumetric flask. Add 14.92 g Triethanolamine (TEA), bring to volume to approximately 950 mL with pure water. Add 1.47 g calcium chloride (CaCl$_2$ · 2H$_2$O) and bring to 1 liter with pure water while adjusting the pH to exactly 7.3 with 6N hydrochloric acid (HCl). The final concentration will be 0.005M DTPA, 0.1M TEA, and 0.01M CaCl$_2$.
(Note: The DTPA reagent should be the acid form.)

6.2 Zinc Standard (1000 mg/L) - Dissolve 1.000 g pure zinc metal in 5-10 mL conc hydrochloric acid (HCl). Evaporate almost to dryness and dilute to 1 liter with Extraction Reagent (see 6.1). Prepare working standards by diluting aliquots of the stock solution standard with Extraction Reagent (see 6.1) to cover the anticipated range in the concentration to be found in the soil extraction filtrate. Working standards from 0.1 to 10 mg Zn/L should be sufficient for most soils.

6.3   Iron Standard (1000 mg/L) - Dissolve 1.000 g pure iron wire in 5-10 mL conc hydrochloric acid (HCl). Evaporate almost to dryness and dilute to 1 liter with Extraction Reagent (see 6.1). Prepare working standards by diluting aliquots of the stock solution standard with Extraction Reagent (see 6.1) to cover the anticipated range in concentration to be found in the soil extraction filtrate. Working standards from 0.1 to 10 mg Fe/L should be sufficient for most soils.

6.4   Manganese Standard (1000 mg/L) - Dissolve 1.582 g manganese oxide MnO_2) in 5 mL conc hydrochloric acid (HCl). Evaporate almost to dryness and dilute to 1 liter with Extraction Reagent (see 6.1). Prepare working standards by diluting aliquots of the stock solution standard with the Extraction Reagent (see 6.1) to cover the anticipated range in concentration to be found in the soil extraction filtrate. Working standards from 0.1 to 10 mg Mn/L should be sufficient for most soils.

6.5   Copper Standard (1000 mg/L) - Dissolve 1.000 g pure copper metal in a minimum amount conc nitric acid (HNO_3) and add 5 mL conc hydrochloric acid (HCl). Evaporate almost to dryness and dilute to 1 liter with Extraction Reagent (see 6.1). Prepare working standards by diluting aliquots of the stock solution with Extraction Reagent (see 6.1) to cover the anticipated range in concentration filtrate. Working standards from 0.1 to 10 mg Cu/L should be sufficient for most soils.

7. PROCEDURE

7.1   Extraction - Weigh 10 g or measure 8.5-cm^3 (see 5.2) of air-dry <10-mesh (2-mm) soil into a 125-mL extraction flask (see 5.3). Add 20 mL Extraction Reagent (see 6.1) and shake on a reciprocating shaker for 2 hours. Samples shaken longer than 2 hours will give high results because a final equilibrium of the metal and soil is not reached in 2 hours. Filter and collect the filtrate.

7.2   Analysis - The elements copper, iron, manganese, and zinc in the filtrate can be determined by either atomic absorption or ICP emission spectrometer.   Since instruments vary in their operating conditions, no specific details are given here. It is recommended that the procedures described by Isaac and Kerber (see 13.2), and Soltanpour et al. (13.6) be followed.

8. CALIBRATION AND STANDARDS

8.1   Working Standards - Working standards should be prepared as described in section 6. If element concentrations are found to be outside the range of the instrument or standards, suitable dilutions should be prepared, starting with a 1:2 soil extract to Extraction Reagent (see 6.1) dilutions.

8.2   Calibration - Calibration procedures vary with instrument techniques and type of instrument. Every precaution should be taken and the manufacturer's recommendations should be followed in the operation and calibration of the instrument.

9. CALCULATION

9.1   The results are reported as kg/ha for a 20 cm depth. Kg element/ha = mg/L of element in extraction filtrate x 4. If the extraction filtrate is diluted, the dilution factor must be applied. For expressing the results in mg/kg of soil, use the following formula:

$$mg/kg \text{ in soil} = mg/L \text{ in solution} \times 2$$

10. EFFECTS OF STORAGE

10.1   Soils may be stored in an air-dry condition for several months with no effects on the amount of copper, iron, manganese, and zinc extracted.

11. INTERPRETATION

11.1   An evaluation of the analysis results as well as accurate fertilizer recommendations for copper, iron, manganese, and zinc must be based on field response for each crop and local field conditions. Interpretative data for critical levels as established by Viets and Lindsay for Colorado soil are available (see 13.3). Boawn did work with DTPA for zinc on Washington soil (see 13.4).

12. COMMENTS

12.1   Grinding can change the amount of DTPA-extractable micronutrients, especially iron. Therefore, it is imperative that grinding procedures be standardized along with extraction procedures. Grinding should be equivalent to using a wooden roller to crush the soil aggregates (see 13.5).

13. REFERENCES

13.1.   Lindsay, W. L., and W. A. Norvell. 1969. Development of a DTPA micronutrient soil test. Agron. Abstracts, p. 84. Equilibrium relationships of Zn^{2+}, Fe^{3+}, Ca^{2+}, and H^+ with EDTA and DTPA in soil. Soil Sci. Soc. Am. Proc. 33:62-68.

13.2   Isaac, R. A., and J. D. Kerber. 1972. Atomic absorption and flame photometry: Techniques and uses in soil, plant and water analysis, pp. 17-24. IN: L. M. Walsh (ed.), Instrumental Methods for Analysis in Soils and Plant Tissue, rev. ed. Soil Science Society of America, Madison, WI.

13.3   Viets, Jr., F. G., and W. L. Lindsay. 1973. Testing soils for zinc, copper, manganese and iron, pp. 153-172. IN: L. N. Walsh and J. D. Beaton (eds.), Soil Testing and Plant Analysis, rev. ed. Soil Science Society of America, Madison, WI.

13.4   Boawn, L. C. 1971. Evaluation of DTPA extractable zinc as a zinc soil test for Washington soils, pp. 143-147. IN: Proceedings of the 22nd Annual Pacific North West Fertilizer Conference, Pacific NW Plant Food Assoc., Bozeman, MT.

13.5   Soltanpour, P. N., A. Khan, and W. L. Lindsay. 1976. Factors affecting DTPA-extractable Zn, Fe, Mn and Cu. Comm. Soil Sci. Plant Anal. 7: 797-821.

13.6   Soltanpour, P. N., J. B. Jones, Jr., and S. M. Workman 1982. Optical emission spectrometry, pp. 29-65. IN: A. L. Page (ed.), Methods of Soil Analysis, Part 2, rev. ed. American Society of Agronomy, Madison, WI.

DETERMINATION OF CADMIUM, COPPER, NICKEL AND ZINC BY DTPA EXTRACTION FOR SLUDGE-AMENDED SOILS

1. PRINCIPLE OF THE METHOD

1.1 The DTPA chelation procedure offers a favorable combination of stability constants for the simultaneous complexing of cadmium, copper, nickel, and zinc (see 1.31). The theoretical basis for the DTPA extraction is the equilibrium of the metal in the soil with the chelating agent. The 7.3 pH, which is buffered with triethanolamine (TEA), prevents excess dissolution of the trace metals.

1.2 The use of DTPA as an extracting agent was developed by Lindsay and Norvell (see 13.2).

2. RANGE AND SENSITIVITY

2.1 Concentration ranges for each element depend on the instrument selected for metal determination. With atomic absorption spectroscopy, cadmium, copper, nickel, and zinc can be determined in soil concentrations of 0.1 to 10 ppm. With plasma emission spectroscopy, the linear range is up to several orders of magnitude greater, depending on the element being determined. Range and upper limits may be extended by diluting the extracting filtrate prior to analysis.

2.2 The sensitivity will vary with the type of instument used and the wavelength selected.

3. INTERFERENCES

3.1 Contamination of soil samples may occur in either the sampling equipment or soil grinding equipment, especially for zinc.

3.2 All apparatus that will come in direct contact with the extraction solution and extraction filtrate must be thoroughly washed and rinsed in pure hydrochloric acid (HCL) and pure water before use. Avoid contact with rubber and metal surfaces.

4. PRECISION AND ACCURACY

4.1 Repeated analysis of the same soil with medium concentration ranges of cadmium, copper, nickel, and zinc will give co-efficients of variability from 10% to 15%. A major portion of the variance is related to the heterogeneity of the soil rather than to the extraction process or method of analysis.

5. APPARATUS

5.1 No. 10 (2-mm opening) sieve.

5.2 8.5-cm^3 scoop, volumetric.

5.3 Extraction flask, 125-mL polyethylene conical flasks.

5.4 Mechanical reciprocating shaker, 180 oscillations per minute.

5.5 Filter funnel, 11 cm.

5.6 Whatman No. 42 ashless filter paper (or equivalent), 11 cm.

5.7 Atomic absorption or plasma emission spectrometer.

5.8 pH meter, line or battery operated, with reproducibility to at least 0.1 pH unit and a glass electrode pair with a calomel reference electrode.

5.9 Analytical balance.

5.10 Volumetric flasks, pipettes and microburet as required for preparation of reagents and standard solutions.

6. REAGENTS

6.1 Extracting Reagent (DTPA diethylenetriaminepentaacetic acid) - Weigh 1.96 DTPA $\{[(HOCOCH_2)_2 \ NCH_2]_2 \ NCH_2COOH\}$ into a 1-liter volumetric flask. Add 14.92 g triethanolamine (TEA). Bring to volume to approximately 950 mL with pure water. Add 1.47 g calcium chloride ($CaCl_2 \cdot 2H_2O$). Bring to 1 liter with pure water while adjusting the pH to exactly 7.3 with 6N HCl. The final concentration will be 0.005M DTPA, 0.1M TEA, and 0.01M $CaCl_2$.
(Note: The DTPA reagent should be the acid form.)

6.2 Cadmium Standard (1000 mg/L) - Dissolve 1.000 g pure cadmium metal in 5 - 10 mL conc hydrochloric acid (HCl). Evaporate almost to dryness and dilute to 1 liter with the Extracting Reagent (see 6.1). Prepare working standards by diluting aliquots of the stock solution standard with the Extracting Reagent (see 6.1) to cover the anticipated range in concentration to be found in the soil extraction filtrate. Working standards from 0.1 to 10 mg Cd/L should be sufficient for most soils.

6.3   Copper Standard (1000 mg/L) - Dissolve 1.000 g pure copper metal in minimum amount conc HNO_3 and add 5 mL conc HCl. Evaporate almost to dryness and dilute to 1 liter with Extracting Reagent (see 6.1). Prepare working standards by diluting aliquots of the stock solution with Extracting Reagent (see 6.1) to cover the anticipated range in concentration filtrate. An initial range of 0 - 100 mg Cu/L is suggested for sludge-amended soils.

6.4   Nickel Standard (1000 mg/L) - Dissolve 1.000 g pure nickel metal in minimum amount conc HNO_3 and add 5 mL conc HCl. Evaporate almost to dryness and dilute to 1 liter with Extracting Reagent (see 6.1). Prepare working standards by diluting aliquots of the stock solution with Extracting Reagent (see 6.1) to cover the anticipated range in concentration filtrate. Working standards from 0.1 to 10 mg Ni/L should be sufficient for most soils.

6.5   Zinc Standard (1000 mg/L) - Dissolve 1.000 g pure zinc metal in 5-10 mL conc HCl. Evaporate almost to dryness and dilute to 1 liter with Extracting Reagent (see 6.1). Prepare working standards by diluting aliquots of the stock solution standard with Extracting Reagent (see 6.1) to cover the anticipated range in concentration to be found in the soil extraction filtrate. An initial range of 0-100 mg Zn/L is suggested for sludge-amended soils.

7. PROCEDURE

7.1   Extraction - Weigh 10 g or measure 8.5-cm^3 (see 5.2) of air-dry <10-mesh (2-mm) soil into a 125-mL extraction flask (see 5.3). Add 20 mL Extracting Reagent (see 6.1) and shake on a reciprocating shaker for 2 hours. Samples that are shaken longer than 2 hours will give high results because a final equilibrium of the metal and soil is not reached in 2 hours. Filter and collect the filtrate.

7.2   Analysis - The elements in the filtrate can be determined by atomic absorption or plasma emission spectroscopy. Since instruments vary in their operating conditions, no specific details are given here. It is recommended that the procedures described by Isaac and Kerber (see 13.2) be followed.

8. CALIBRATION AND STANDARDS

8.1   Working Standards - Working standards should be prepared as described in section 6. If element concentrations are found outside the range of the instrument or standards, suitable dilutions should be prepared starting with a 1:2 soil extract to Extracting Reagent (see 6.1) dilutions.

8.2   Calibration - Calibration procedures vary with instrument techniques and type of instrument. Every precaution should be taken to ensure that the proper procedures are taken and the manufacturer's recommendations followed in the operation and calibration of the instrument used.

9. CALCULATION

9.1   The results are reported as kg/ha for a 20-cm depth. Kg of element/ ha = mg/L of element in extraction filtrate x 4. If the extraction filtrate is diluted, the dilution factor must be applied. To express the results in mg/kg of soil, use the following formula:

mg/kg in soil = mg/L in solution x 2

10. EFFECTS OF STORAGE

10.1   Soils may be stored in an air-dry condition for several months with no effects on the amount of zinc, iron, manganese, and copper extracted.

11. INTERPRETATION

11.1   Limited work has been done to evaluate soil test methods for sludge-amended soils. Rappaport et al. (13.4, 13.5) reported that the DTPA method correlated well with metals applied in sludge but found generally poor correlations with plant uptake. More research is needed in this area, particularly with sensitive crops.

12. COMMENTS

12.1.   Grinding can change the amount of DTPA - extractable heavy metals, especially iron. Therefore, it is imperative that both grinding procedures and extraction procedures be standardized. The grinding process should approximate using a wooden roller to crush the soil aggregates (see 13.5).

13. REFERENCES

13.1   Baker, D. E., and M. C. Amacher. 1982. Nickel, copper, zinc, and cadmium, pp. 323-336. IN: A. L. Page (ed.), Methods of Soil Analysis, Part 2, Agronomy 9. American Society of Agronomy, Madison, WI.

13.2   Lindsay, W. L., and W. A. Norvell. 1969. Development of a DTPA micronutrient soil test. Agron. Abstracts, p. 84. Also found in Equilibrium relationships of Zn^{2+}, Fe^{3+},Ca^{2+}, and H^+ with EDTA and DTPA in soil. Soil Sci. Soc. Am. Proc. 33:62-68.

13.3   Isaac, R. A., and J. D. Kerber. 1972. Atomic absorption and flame photometry: Techniques and uses in soil, plant, and water analysis, pp. 17-24. IN: L. M. Walsh, (ed.), Instrumental Methods for Analysis of Soils and Plant Tissue, rev. ed. Soil Science Society of America, Madison, WI.

13.4   Rappaport, B. D., J. D. Scott, D. C. Martens, R. B. Reneau, Jr., and T. W. Simpson. 1987. Availability and distribution of heavy metals, nitrogen, and phosphorus from sewage sludge in the plant-soil-water continuum. VPI-VWRRC-BULL 154 5C. Virginia Water Resources Research Center, Blacksburg, VA.

13.5   Rappaport, B. D., D. C. Martens, R. B. Reneau, Jr., and T. W. Simpson. 1988. Metal availability in sludge-amended soils with elevated metal levels. J. Environ. Qual. 17:42-47.

13.6   Soltanpour, P. N., A. Khan, and W. L. Lindsay. 1976. Factors affecting DTPA-extractable Zn, Fe, Mn and Cu. Comm. Soil Sci. Plant Anal. 7:797-821.

NITRATE-NITROGEN DETERMINATION BY SPECIFIC-ION ELECTRODE

1. PRINCIPLE OF THE METHOD

1.1   In this procedure, a digital pH/mV meter and a specific ion (NO_3) electrode with a double junction reference electrode are used to determine available (soluble) nitrate-nitrogen (NO_3-N). The technique and principles of the procedure are similar to those involved in measuring pH with a glass electrode. The NO_3 is measured in a suspension of soil with water or a dilute salt such as ammonium sulfate [$(NH_4)_2SO_4$] or calcium sulfate [$CaSO_4$] and is usually reported as NO_3-N. Nitrogen so measured is often called "residual" or "carryover" N and is subtracted from the total crop N requirement as a final step in determining fertilizer N requirement. This measurement is most useful in cropping systems in which the soil is not highly leached. However, the measurement can be used even in leached soils when samplings are properly timed and include a subsoil sample (Ward 1971).

1.2   The use of this method, as well as general interest in the NO_3-N soil testing, dates back to the late 1960s (Ward 1971). Early research demonstrated that the procedure was relatively free of interference from other ions commonly found in soils (Oien and Selmer-Olsen 1969, Potterton and Shults 1967). A major disadvantage of using the early electrodes was the high level of skill and dexterity required to replace and reassemble the electrode every two weeks. The newer electrode has a screw-on nodule which eliminates this problem.

2. RANGE AND SENSITIVITY

2.1   The electrode operates linearly in the range of about 1 to 1,000 μg/mL NO_3-N.

2.2   Sensitivity of the electrode procedure in soil testing is a function of the soil-to-solution ratio. The technique may be used even for a crop with a moderate N requirement, such as cotton. A soil-solution ratio of 1:2.5 will allow measurement of levels as low as 5 lb/A.

3. **INTERFERENCES**

3.1 Of the ions that are commonly found in soil extracts, chloride (Cl), carbonate (CO_3), and bicarbonate (HCO_3) are the most likely to cause interference. However, even these ions have only a negligible effect. At concentrations of up to 400 μg/mL, these ions will cause less than a 10% error in analysis of 10 μg/mL NO_3-N. Specific details of interferences, as well as electrode principles and operational procedures, are usually provided with the electrode purchase (Orion Research 1981).

4. **PRECISION AND ACCURACY**

4.1 Repeated analyses of different control samples used over several months in the Oklahoma State University Agronomic Services Laboratory have demonstrated a reproducibility of ± 2 to 3 lbs/A of the mean value of samples containing 10 to 50 lbs/A of NO_3-N. When reruns of the test are requested, results, even on soils containing over 100 lbs/A of NO_3-N, usually fall within ± 10% of the initial test value.

5. **APPARATUS**

5.1 No. 10 (2-mm opening) sieve.

5.2 8.5 cm^3 scoop, volumetric (10 g).

5.3 100-mL extraction container. Disposable 3.5-oz plastic Solo cups, fitted into a styrofoam base, are inexpensive and work very well.

5.4 A digital pH/mV meter.

5.5 A double junction reference electrode. Use the calcium sulfate extraction solution (below) as the outer chamber filling solution instead of the filling solution shipped with the reference electrode. Replace the solution in the outer chamber daily.

5.6 A nitrate-ion (NO_3) sensitive electrode.

5.7 A magnetic stir bar and plate (or other suitable means for gently suspending soil in small samples).

6. REAGENTS

6.1 Extracting Reagent - Dissolve 2.0 g calcium sulfate ($CaSO_4$) or 2.53 gm calcium sulfate dihydrate ($CaSO_4 \cdot 5H_2O$)in pure water and bring to 1.0 liter with pure water.

6.2 Primary Standard Solution - Weigh 7.220 g dry potassium nitrate (KNO_3) into a 1-liter volumetric flask. Dissolve the potassium nitrate and bring to 1-liter volume with calcium sulfate extracting solution (see 6.1). Label 1000 µg/mL N as potassium nitrate.

6.3 Working Standard - Prepare a 100 µg/mL working standard solution by transferring 100 mL primary standard (see 6.2) to a 1-liter volumetric flask and bringing to volume with extracting solution. Both the primary standard and the working standard solutions should be stored in the refrigerator.

6.4 Calibration Standards - Prepared calibration standards using the working standard solution (see 6.3) according to the following table:

MLS Working Standard	Actual µg/mL NO_3-N	Readings in lbs N/A
0	0	0
35	5	25
50	10	50
100	20	100

7. PROCEDURE

7.1 Measure 10 g air-dry, sieved soil (10 mesh, see 5.1) into the extraction container. Add 25 mL extracting solution (see 6.1) and shake for 30 minutes on a rotary shaker at 150 rotations/minute.

7.2 Measure NO_3-N in soil/extraction solution suspension by immersing the electrode directly into the suspension while it is being stirred. Analyze a standard solution every 10 samples and restandardize if necessary.

8. CALIBRATION AND STANDARDS

8.1 <u>Working Standards</u> - Working standards should be prepared as described in section 6.4. When they are stored in a refrigerator, these solutions may be prepared in sufficient volume to satisfy laboratory needs for one week.

8.2 Standardize the instrument using the 100 lbs N/A solution for the high standard and the 25 lbs N/A solution for the low standard with a two-point standardization. Solutions should be stirred while readings are being taken. After each determination, the electrode should be rinsed with distilled water and blotted.

Verify reliable (linear) standardization by analyzing the remaining calibration solutions (see 6.4).

9. CALCULATIONS

9.1 The results are reported as lbs NO_3-N per acre furrow slice at a depth of 62/3 inches. The same testing procedure is used to measure available NO_3-N in the subsoil. The surface and subsoil values may be combined after adjustments to the subsoil value have been made to account for the difference in the soil depth sampled.

10. EFFECTS OF STORAGE

10.1 Soils may be stored in an air-dry condition for several months without affecting the available NO_3-N contents.

10.2 After extraction, the suspensions should be analyzed as soon as possible. Results are unlikely to change significantly within 12 hours after extraction.

11. INTERPRETATION

11.1 Evaluation of the results with respect to the adequacy of N for crop production is possible when the N requirement of different crop yield goals has been established. In the simplest, most direct interpretation, theN fertilizer requirement is calculated by subtracting the soil test N value from the N requirement from crop yield goal (see 12.5).

12. REFERENCES

12.1 Ward, R. C. 1971. NO_3-N soil test: Approaches to use and interpretation. Commun. Soil Sci. Plant Anal. 2(2):61-71.

12.2   Oien, A. and A. R. Selmer-Olsen. 1969. Nitrate determination in soil extracts with the nitrate electrode. Analyst 94:888-894.

12.3   Potterton, W. W. and W. D. Shultz. 1967. An evaluation of the performance of the nitrate-selective electrode. Analytical Letters 1(2):1I-22.

12.4   Orion Research. 1981. Nitrate Ion Electrode Model 93-07 Instruction Manual. Orion Research, Inc., Cambridge, MA.

12.5   Johnson, G. V. and B. B. Tucker. 1982. OSU soil test calibrations. Oklahoma State University Extension, Facts No. 2225. Oklahoma State University Cooperative Extension Service, Stillwater, OK 74078.

DETERMINATION OF POTASSIUM, CALCIUM, MAGNESIUM AND SODIUM BY WATER EXTRACTION

1. PRINCIPLE OF THE METHOD

1.1   This method uses water to extract potassium, calcium, magnesium, and sodium from soil. A soil-water ratio of 1:5 (v:v) is the one adapted for routine analysis. (see 12.1, 12.2 and 12.3).

1.2   This method is relatively simple and can serve for quick and routine scanning. However, it suffers from the setback of unrealistic figures for calcium and sodium due to cation exchange equilibrium shift (see 12.3).

2. RANGE AND SENSITIVITY

2.1   The range of detection will depend on the particular instrument setup. The range can be extended by diluting the extract.

2.2   The sensitivity will vary with the type of instrument used, wavelength selected, and method of excitation or dissociation.

2.3   The commonly used methods of analysis are flame emission and atomic absorption spectroscopy. A more complete description of these methods is given by Isaac and Kerber (see 12.4) and Christian (see 12.5).

3. INTERFERENCES

3.1   A number of cations and anions which are extracted along with the desired cations will interfere with their determination, particularly calcium by atomic absorption spectroscopy.

3.2   Known interferences must be eliminated by adding certain cations such as lanthanum, depending on the flame condition used (see 12.5).

4. PRECISION AND ACCURACY

4.1   Repeat analysis of the same soil in the medium concentration range of potassium, calcium, magnesium, and sodium will give coefficients of variation of 5% to 10%. A major portion of the variance is related to the heterogeneity of the soil rather than to the extraction or method of analysis.

5. APPARATUS

5.1 No. 10 (2-mm opening) sieve.

5.2 4.25-cm^3 scoop, volumetric.

5.3 Extraction bottle or flask, 50 mL, with stoppers.

5.4 Mechanical reciprocating shaker, 180 oscillations per minute.

5.5 Filter funnel, 11 cm.

5.6 Whatman No. 1 filter paper (or equivalent), 11 cm.

5.7 Atomic absorption and/or flame emission spectrophotometer.

5.8 Funnel racks.

5.9 Analytical balance.

5.10 Volumetric flasks and pipettes as required for preparation of reagents and standard solutions.

6. REAGENTS

6.1 Extraction Reagent - Pure water.

6.2 Potassium Standard (1000 mg/L) - Weigh 1.9080 g potassium chloride (KCl) into a 1-liter volumetric flask and bring to volume with pure water. Prepare working standards by diluting aliquots of the stock solution standard to cover the anticipated range in concentration to be found in the soil extraction filtrate. Working standards from 5 to 100 mg K/L should be sufficient for most soils.

6.3 Calcium Standard (1000 mg/L) - Weigh 2.498 g calcium carbonate (CaCO$_3$) into a 1-liter volumetric flask, add 50 mL of pure water and add dropwise a minimum volume of conc HCl (approximately 20 mL) to effect complete solution of the calcium carbonate. Dilute to the mark with pure water. Prepare working standards by diluting aliquots of the stock solution standard with pure water to cover the anticipated range in concentration to be found in the soil extraction filtrate. Working standards should be prepared based on the expected concentration in the extract and the range of analytical procedure employed.

6.4 Sodium Standard (1000 mg/L) - Weigh 2.542 g sodium chloride (NaCl) into a 1-liter volumetric flask and bring to volume with pure water. Prepare the working standards by diluting aliquots of the stock solution standard with pure water to cover the anticipated range in concentration to be found in the soil extraction filtrate. Working standards from 1 to 10 mg Na/L should be sufficient for most soils.

6.5   Magnesium Standard (1000 mg/L) - Weigh 1.000 g magnesium ribbon into a 1-liter volumetric flask and dissolve in the minimum volume of dilute HCl and bring to volume with pure water. Prepare the working standards by diluting aliquots of the stock solution standard with pure water to cover the anticipated range in concentration to be found in the soil extraction filtrate. Working standards from 5 to 50 mg Mg /L should be sufficient for most soils.

7. PROCEDURE

7.1   Extraction - Weigh 5 g or scoop 4.25-cm (see 5.2) of air-dry, <10 mesh (2 mm) soil into a 50-mL extraction bottle (see 5.3). Add 25 mL pure water. Seal the bottle with a stopper and shake for 30 minutes on a reciprocating shaker (see 5.4). Allow to stand for 15 minutes to let the bulk of soil settle. Filter the supernatant liquid. Discard the initial filtrate if it is turbid.

7.2   Analysis - Calcium and magnesium in the filtrate can be determined by atomic absorption spectroscopy using either an air-acetylene or a nitrous oxide-acetylene flame. Potassium and sodium are determined by flame emission. Since instruments vary in their operating conditions, no specific details are given here. However, procedures recommended by the manufacturer and described in the operation manual should be followed.

8. CALIBRATION AND STANDARDS

8.1   Working Standards - Working standards should be prepared as described in section 6. If the element concentrations in the extract are found to be outside the range of the instrument or standards, suitable dilutions should be prepared. Dilution should be made to minimize magnification of the error introduced by diluting.

8.2   Calibration - Calibration procedures vary with instrument techniques and the type of instrument. Every precaution should be taken to ensure that the proper procedures are followed, and the manufacturer's recommendations in the operation and calibration of the instrument should be followed.

9. CALCULATION

9.1   The results are reported as kg/ha for a 20 cm depth of soil. Kg of element/ha = mg/L of element in extraction filtrate x 10.

9.2   To convert to other units for comparison, see Mehlich (12.6).

10. EFFECTS OF STORAGE

10.1   Soils may be stored in an air-dry condition for several months with no effect on element content.

10.2   After extraction, the filtrate should not be stored any longer than 24 hours unless it is treated with acid to prevent bacterial growth and the precipitation of some elements (particularly calcium).

11. INTERPRETATION

11.1   An evaluation of the analytical results, as well as accurate fertilizer recommendations, must be based on field response data conducted under local soil-climate-crop conditions (see 12.7).

12. REFERENCES

12.1.   Bower, C. A., and L. V. Wilcox. 1965. Soluble salts, pp. 935-945. IN: C. A. Black (ed.), Methods of Soil Analysis, Part 2. Agronomy No. 9. American Society of Agronomy, Madison, WI.

12.2.   Hesse, P. R. 1971. A Textbook of Soil Chemical Analysis. Chemical Publishing Co., New York.

12.3.   Chapman, H. D., and P. Pratt. 1982. Method of Analysis for Soils, Plants and Water, priced publication 4034. University of California, Division of Agricultural Sciences, Berkeley, CA.

12.4.   Isaac, R. A., and J. D. Kerber. 1971. Atomic absorption and flame photometry: Techniques and uses in soil, plant and water analysis, pp. 17-38. IN: L. M. Walsh (ed.), Instrumental Methods for Analysis of Soils and Plant Tissue. Soil Science Society of America, Madison, WI.

12.5   Christian, G. D. 1970. Atomic Absorption Spectroscopy. Wiley Interscience, New York.

12.6   Mehlich, A. 1972. Uniformity of expressing soil test results: A case for calculating results on a volume basis. Comm. Soil Sci. Plant Anal. 3:417-424.

12.7 Doll, E. C., and R. E. Lucas. 1973. Testing soils for potassium, calcium and magnesium, pp. 133-152. IN: L. M. Walsh and J. D. Beaton (eds.), Soil Testing and Plant Analysis. Soil Science Society of America, Madison, WI.

DETERMINATION OF ORGANIC MATTER BY WET DIGESTION

1. PRINCIPLE OF THE METHOD

1.1 The total soil organic matter is routinely estimated by measuring organic carbon content. The procedure is described by Mebius (see 12.1).

1.2 The method described is a wet-oxidation procedure using potassium dichromate with external heat and back titration to measure the amount of unreacted dichromate. This method and other methods are thoroughly discussed by Hesse (see 12.2), Jackson (see 12.3) and Allison (see 12.4). The procedure is rapid and adapted for routine analysis in a soil testing laboratory. It is primarily used to determine the organic matter of mineral soils.

2. RANGE AND SENSITIVITY

2.1 The method is useful for soils containing very low organic carbon to as high as 12% organic carbon with a sensitivity of about 0.2% to 0.5% organic carbon.

3. INTERFERENCES

3.1 Soils containing large quantities of chloride (Cl^-), manganous (Mn^{++}) and ferrous (Fe^{++}) ions will give high results. The chloride (Cl^-) interference can be eliminated by adding silver sulfate (Ag_2SO_4) to the Oxidizing Reagent (see 6.1). No known procedure is available to compensate for the other interferences.

3.2 The presence of $CaCO_3$ up to 50% causes no interferences.

3.3 This procedure is not recommended for high organic matter content soils or organic soils (see 2.1).

4. PRECISION AND ACCURACY

4.1 The method is an incomplete digestion and a correction factor must be applied. The correction factor used is 1.15 (see 12.4).

4.2 Repeated analyses should give results with a coefficient of variability of no greater than 1% to 4%.

4.3 Soil samples must be thoroughly ground and mixed before subsampling because heterogeneity is a serious problem in organic matter distribution within samples.

TITRATION   PROCEDURE

5 . APPARATUS

5.1   No. 10 (2-mm opening) sieve.

5.2   500-mL Erlenmeyer flasks with ground glass tops.

5.3   Reflux condenser apparatus and hot plate.

5.4   Titration apparatus and burette.

5.5   Glassware and pipettes for dispensing and preparing reagents.

5.6   Analytical balance.

6 . REAGENTS

6.1   0.267N Potassium Dichromate - Dissolve 13.072 g potassium dichromate $(K_2Cr_2O_7)$ in 400 mL pure water. Add 550 mL conc sulfuric acid (H_2SO_4). Let it cool and dilute to 1 liter with pure water.

6.2   0.2M Mohr's Salt Solution - Dissolve 78.390 g ferrous ammonium sulfate $[Fe\ (NH_4)_2\ (SO_4) \cdot 6H_2O]$ in 500 mL pure water and add 50 mL conc sulfuric acid (H_2SO_4) . Let it cool and dilute to 1 liter. Prepare fresh for each use.

6.3   Indicator Solution - Dissolve 200 mg n-phenylanthranilic acid in 1 liter of a 0.2% sodium cabonate (Na_2CO_3) solution.

7. PROCEDURE

7.1   Weigh 0.1 to 0.5 g (depending on estimated organic content) of air-dry, <10-mesh (2-mm) soil into a 500-mL Erlenmeyer flask. Add 15 mL 0.267N potassium dichromate (see 6.1). Connect the flask to the reflux condenser and boil for 30 minutes. Let cool. Wash down the condenser and flush with pure water. Add 3 drops of the Indicator Solution (see 6.3). Titrate with Mohr's Salt Solution (see 6.3) at room temperature. As the end point is approached, add a few more drops of the Indicator Solution. The color change is from violet to bright green.

8. CALIBRATION AND STANDARDS

8.1   Reagents 6.1 and 6.2 should be carefully prepared.

8.2   A blank analysis is carried through the procedure where no soil is added.

9. CALCULATION

9.1   % organic carbon = $\dfrac{(\text{meq } K_2Cr_2O_7 - \text{meq } FeSO_4) \times 0.3}{\text{grams soil}}$ x 1.15

9.2   % organic matter = % organic carbon x 1.724.

10. EFFECTS OF STORAGE

10.1   Air dry soil may be stored many months in closed containers without affecting the organic matter content of the soil.

11. INTERPRETATION

11.1   The interpretation of the organic matter content of a soil will depend on the use of this parameter. Organic matter content can be used to evaluate N availability and herbicide effectiveness reaction.

12. REFERENCES

12.1.   Mebius, L. J. 1960. A rapid method for the determination of organic carbon in soil. Anal. Chem. Acta. 22:120-124.

12.2.   Hesse, P. R. 1971. Soil Chemical Analysis. Chemical Publishing Co., New York.

12.3.   Jackson, M. L. 1958. Soil Chemical Analysis. Prentice-Hall, Englewood Cliffs, NJ.

12.4.   Allison, L. E. 1965. Organic carbon, pp. 1367-1378. IN: C. A. Black (ed.), Methods of Soil Analysis, Part 2, Agronomy No. 9. American Society of Agronomy, Madison, WI.

COLORIMETRIC PROCEDURE

5. APPARATUS

   5.1   No. 10 (2-mm opening) sieve.

   5.2   1.5-cm^3 scoop, volumetric.

   5.3   200-mL test tube.

   5.4   Delivery burette, or 20-mL automatic pipette.

   5.5   Colorimeter (or spectrophotometer) for measuring absorbance (A) at 645 nm (red filter).

   5.6   Analytical balance.

   5.7   Volumetric flasks and pipettes as required for preparation of reagents and standard solutions.

6 . REAGENTS

   6.1   0.67M Sodium dichromate - Dissolve 4.000 g sodium dichromate $(Na_2Cr_2O_7)$ in pure water and dilute to 20 liters with pure water.

   6.2   Technical grade sulfuric acid (H_2SO_4).

7 . PROCEDURE

   7.1   Weigh 2.0 g or volume scoop 1.5-cm^3 (see 5.2) of air-dry, <10-mesh (2-mm) soil into a 200-mL test tube (see 5.3). Under a hood, add 20 mL 0.67 M sodium dichromate (see 6.1) and then add 20 mL sulfuric acid (see 6.2). Mix thoroughly (with CAUTION) and allow it to cool at least 40 minutes. After cooling, add 100 mL pure water, mix and allow it to stand at least 8 hours. An aliquot of the clarified solution is transferred to a colorimeter vial using a syringe pipette. Measure absorbance (A) at 645 nm (see 5.5).

8. CALIBRATION AND STANDARDS

   8.1   A standard curve is established with several soils that have an adequate range of organic matter contents. The percentage of organic matter is determined by a standard method (see Titration Procedure). Absorbance values are determined for each known soil organic matter and a curve is constructed by plotting the percentage of organic matter versus absorbance, including a reference sample with daily runs of the method aids in verifying equivalent conditions between the standard curve and daily runs.

9. EFFECTS OF STORAGE

9.1   Soils may be stored in an air-dry condition with no effect on the percentage of organic matter.

10. REFERENCES

10.1   Jackson, M. L. 1958. Soil Chemical Analysis. Prentice-Hall, Inc., Englewood Cliffs, NJ.

10.2   Schollenberger, C. J. 1927. A rapid approximate method for determining soil organic matter. Soil Sci. 24:65-68.

10.3   Walkley, Allen. 1947. A critical examination of a rapid method for determination of organic carbon in soils: Effect of variation in digestion conditions and of inorganic soil constituents. Soil Science 63:251-257.

10.4   Graham, E. R. 1948. Determination of soil organic matter by means of a photoelectric colorimeter. Soil Science 65:181-183.

DETERMINATION OF ORGANIC MATTER BY LOSS-ON-IGNITION

1. PRINCIPLE OF THE METHOD

1.1 Total soil organic matter is estimated by loss-on-ignition (LOI). The procedure was initially described by Davies (see 9.1), and the method described here is given by Ben-Dor and Banin (see 9.2).

1.2 The method described is a procedure in which a soil sample is dried at 105°C and then ashed at 400°C. The loss in weight between 105°C and 400°C constituents the organic matter content. The results obtained compare favorably with those obtained by the dichromate wet-oxidation method (see page 167) and by carbon analyzers (see 9.2, 9.3, 9.4, 9.5, 9.6). Others have used different ashing temperatures ranging from 360°C (see 9.4) to 600°C (see 9.2, 9.5).

2. RANGE AND SENSITIVITY

2.1 The method is useful for soils containing low to very high organic matter contents with a sensitivity of about 0.2% to 0.5% organic matter.

3. INTERFERENCES

3.1 The method is generally considered not suitable for organic matter determination for calcareous soils. The presence of $CaCO_3$ may interfer.

4. PRECISION AND ACCURACY

4.1 Consistent analytical results are obtainable under a range of sample sizes, ashing vessels, and ashing temperatures and length of ashing time. However, mineral composition of the soil may be a factor in the determination and may require more than one calibration curve (see 9.4). In addition, soil horizons may be another factor affecting LOI results (see 9.7).

4.2 An automated system for determining organic matter content by LOI has been described by Storer (see 9.8).

4.3 Repeated analyses should give results with a coefficient of variability of no greater than 1% to 4%.

5. APPARATUS

5.1 No. 10 (2-mm opening) sieve.

5.2   50-mL beaker or other suitable ashing vessel.

5.3   Drying oven.

5.4   Muffle furnace.

5.5   Analytical balance (0.01 g sensitivity).

6. PROCEDURE

6.1   Weight 5.00 to 10.00 g sieved (see 5.1) soil into an ashing vessel (see 5.2).

6.2   Place the ashing vessel + soil into the drying oven (see 5.3) set at 105°C and dry for 4 hours.

6.3   Remove the ashing vessel from the drying oven and place in a dry atmosphere. Once cool, weigh to the nearest 0.01 g.

6.4   Place the ashing vessel + soil into a muffle furnce (see 5.4) and bring the temperature to 400°C. Ash in the furnace at 400°C for 4 hours.

6.5   Remove the ashing vessel from the muffle furnace, let cool in a dry atmosphere and weigh to the nearest 0.01 g.

7. CALIBRATION AND STANDARDS

7.1   The percent organic matter in the soil is determined by the formula: % OM = [W105 - W400) x 100]/W105,

where: W105 is weight of soil at 105°C and W400 is weight of soil at 400°C.

7.2   A standard curve may be established with several soils that have a range of organic matter contents encompassing that in the unknowns. The percentage of organic matter in the standards will have been determined by other methods. More than one calibration curve may be required for varying soil types (see 9.4 and 9.7).

8. EFFECTS OF STORAGE

8.1   Soils may be stored in an air-dry condition with no effect on the percentage of organic matter.

9. REFERENCES

9.1   Davies, B. E. 1974. Loss-on-ignition as an estimate of soil organic matter. Soil Sci. Soc. Amer. Proc. 38:150-151.

9.2   Ben-Dor, E. and A. Banin. 1989. Determination of organic matter content in arid-zone soils using a simple "loss-on-ignition" method. Commun. Soil Sci. Plant Anal. 20(15-16): 1675-1695.

9.3   Golden, A. 1987. Reassessing the use of loss-on-ignition for estimating organic matter content in noncalcareous soils. Commun. Soil Sci. Plant Anal. 18(9):1111-1116.

9.4   Schulte, E. E., C. Kaufmann, and J. B. Peter. 1991. The influence of sample size and heating time on soil weight loss-on-ignition. Commun. Soil Sci. Plant Anal. 22(1-2):159-168.

9.5   Gallardo, J. F. and J. Saavedra. 1987. Soil organic matter determination. Commun. Soil Sci. Plant Anal. 18(6):699-707.

9.6   Lowther, J. R., P. J. Smethurst, J. C. Carlyle, and E. K. S. Mabiar. 1990. Methods for determining organic carbon in Podzolic sands. Commun. Soil Sci. Plant Anal. 21(5-6):457-470.

9.7   David, M. B. 1988. Use of loss-on-ignition to assess soil organic carbon in forest soils. Commun. Soil Sci. Plant Anal. 19:1593-1599.

9.8   Storer, D. A. 1984. A simple high sample volume ashing procedure for determination of soil organic matter. Commun. Soil Sci. Plant Anal. 15(7):759-772.

COLORIMETRIC DETERMINATION OF HUMIC MATTER WITH 0.2N NaOH EXTRACTION

1. PRINCIPLE OF THE METHOD

1.1   This extraction method is designed to determine the sodium hydroxide soluble humic matter, which consists of humic and fulvic acids. These components comprise approximately 85% to 90% of the soil humus and are responsible for the cation and anion exchange properties exhibited by the soil organic fraction.

1.2   This method is based on the concept that humic matter compounds are soluble in dilute alkali solutions (see 6.1, 6.2 and 6.5). Acidic organic compounds are converted to ions with the subsequent formation of a physical solution of these ions in water (see 6.5). The reaction of a dilute alkali with the humic matter results in a colored solution which is proportional to the soluble humic matter content within the soil. The color varies from shades of brown to black, depending on the type of soil from which the sample originates. Colorimetric determination of the humic matter content of soils by this method is based upon the color intensity of the solution following extraction with a dilute alkali extractant. The alkali used in the method is NaOH, which serves as the humic acid solvent. DTPA aids in the dispersion of some of the large molecular Ca-humate compounds, and ethanol aids in the solubility of hydrophobic lipid components of soil organic matter. Calibration data was generated from a standard humic matter source (see 4.5).

1.3   This method was designed to accomplish two major objectives: (1) to estimate the chemically reactive portion of the soil organic fraction for better prediction of herbicide rate requirements and (2) to remove chromium from the effluent of municipal waste systems. Experimental evidence has shown that this method can be used to predict herbicide rates (see 6.3 and 6.4).

2. RANGE AND SENSITIVITY

2.1   Up to 10% of the humic matter content of soils can be determined by this method. Higher levels could be determined with a wider extraction ratio (see 6.6). The method as described will encompass a majority of mineral soils. Saturation of the method is encountered on the organic and mineral organic soils where total organic matter is high. However, in some organic soils, the humic matter content is low even though the percentage of combustible organic content may be in excess of 90%.

2.2 The sensitivity of this method would depend on the quality and homogeneity of the field sample.

3. APPARATUS

3.1 No. 10 (2-mm opening) sieve.

3.2 1.0 cm^3 (volumetric) soil measure and teflon-coated leveling rod.

3.3 55-mL polystyrene extraction vials (35 mm D x 75 mm H).

3.4 Automatic dispenser for extractant, 20-mL capacity.

3.5 Diluter-dispenser, 5:35-mL capacity.

3.6 Analytical balance.

3.7 Photometric colorimeter suitable for measuring in the 650-nm range. Colorimeters equipped with moveable fiber optic probes can be used to read samples directly from diluted sample vials.

4. REAGENTS

4.1 All reagents are ACS analytical grade unless otherwise stated.

4.2 Sodium hydroxide (NaOH).

4.3 DTPA (Diethylenetriaminepenta acetic acid, pentasodium salt) - Technical grade (40 $\pm$ 1% in H_2O), fw = 503.26. Density = 1.26.

4.4 Ethyl alcohol, denatured (C_2H_5OH).

4.5 Standard humic acid (Aldrich Chemical Co., 940 W. St. Paul Ave., Milwaukee, WI 53233).

4.6 Extracting Solution 0.2N NaOH - 0.0032M DTPA - 2% Alcohol. Using a 4-1iter volumetric flask, add about 1000 mL pure water, 32 g NaOH (see 4.2) and dissolve. Then add 16 mL DTPA (see 4.3) and 80 mL ethanol (see 4.4). Make to volume with pure water and mix thoroughly. Larger volumes of extractant can be prepared, depending on the numbers of samples to be analyzed.

5. PROCEDURE

5.1 Standard Humic Matter Calibration - Dry the humic acid standard (see 4.5) at $105°$ C for approximately 4 hours. Loss on ignition at $550°$ C shows that this humic matter standard contains 87% organic matter. For calibration, weigh 0.115 g standard humic acid ($0.1 \div 0.87 = 0.115$) and place into a 55 cm^3 polystyrene vial (see 3.3). Add 20 mL extractant (see 4.6) with sufficient force to mix the sample.

Allow the sample to set for 1 hour and then add an additional 20 mL extractant (see 4.6) with sufficient force to mix well. The two 20-mL portions of extractant are added separately to enhance dissolution of the humic matter. Let the sample set overnight (16-18 hours minimum), then pipette 5 mL of the supernatant and 35 mL water into 55-mL polystyrene vials (see 3.3). (Caution should be taken not to pipette colloidal precipitation from the bottom of vials.)

The final dilution of the sample is a 1:8 ratio (5 mL sample + 35 mL water), which is required at this extraction ratio to get within the instrument reading range. Set the instrument at 100% T with 5 mL extractant and 35 mL water. Read the standard at 650 nm. When a Brinkman probe colorimeter with a 2-cm light path is used, the standard humic acid standard should read 10% T. This equates to 10 $g/100$ cm^3 humic matter equivalent. A standard curve can be developed by sequential 1:1 dilutions of the 10% humic acid standard. To develop the factor for converting the instrument reading to g HM/100 cm^3, convert %T to absorbance, then divide g HM/100 cm^3 by the absorbance. Assuming linearity of the standard, the ratio of g HM/absorbance should be a constant.

If a larger volume of humic acid standard is required for calibration, multiple quantities of standard humic acid and extractant can be used.

5.2 Soil Sample Extraction and Analysis - Measure 1 cm^3 soil (screened-2mm) into 55-mL polystyrene vials (see 3.3) and add 20 mL alkali extractant (see 4.6) with sufficient force to mix well. After 1 hour, add another 20 mL extractant (see 4.6) with mixing force and allow the samples to set overnight. In addition to allowing adequate reaction time of humic matter with the extractant, setting allows soil particles to settle out, leaving a clear supernatant. Transfer 5 mL of undisturbed supernatant and 35 mL of water into 55-mL polystyrene vials. Set the instrument to read 100% T with a blank (5 mL extractant + 35 mL water). Read the samples at 650 nm and record the %T. A check sample whose humic matter content has been previously determined should be analyzed routinely with unknown samples. Samples which exceed 10% HM can be diluted with water and the appropriate dilution factor employed.

5.3   Calculations - The humic matter (HM) content of a soil can be determined from a standard curve or by converting %T to Abs and multiplying by the factor developed in the calibration procedure (see 5.5). For this method the factor is 10, therefore Abs x 10 = gm HM equiv/100 cm^3 of soil. If the percentage of HM on a weight basis is desired, divide HM (g/100 cm^3) by the WV (weight/volume in g/cm^3) of each soil. For specific values in the developent of this procedure, see 5.4.

5.4   Calibration Procedure - The values shown below were developed to determine HM up to 10%, using an extraction ratio of 1:40 (1 cm^3 soil + 40 mL extractant, see 5.1), with 0.115 g of humic acid standard (87% organic matter).

HM Equiv,[1] g/100 cm	Abs	HM Equiv/Abs	Factor[3]
10.000	–	–	
5.000	–	–	
2.500	–	–	
1.250[2]	1.000	1.25	10
0.625	0.509	1.23	
0.313	0.206	1.21	
0.156	0.131	1.19	
0.078	0.061	1.28	
0.036	0.027	<u>1.33</u>	
		Avg 1.25 x 8 = 10	

1/   The Standard HA sample is diluted sequentially (1:1) with H_2O for the development of the standard curve.

2/   The Standard HA sample is diluted 1:8 (5 cm^3 sample extract + 35 mL H_2O). Read at 650 nm = 10% T or 1.0% Abs. Unknown samples can be diluted in the same manner.

3/   The factor is determined by taking the average of HM/abs multiplied by 8 (d.f.)

6. REFERENCES

6.1   Hayes, M. H. B., R. S. Swift, R. E. Wardle, and J. K. Brown. 1975. Humic materials from an organic soil: A comparison of extractants and properties of extracts. Geoderma 13:231-245.

6.2   Levesque, M., and M. Schnitzer. 1967. The extraction of soil organic matter by base and chelating resin. Can. J. Soil Sci. 47:76-78.

6.3   Strek, H. J., and J. B. Weber. 1983. Update of soil testing and herbicide rate recommendations. Proc. South. Weed Sci. Soc. 36:398-403.

6.4   Weber, J. B., and C. John Peter. 1982. Absorption, bioactivity, and evaluation of soil tests for Alachlor, Acetochlor and Metolochlor. Weed Sci. 30:14-20.

6.5   Mortensen, J. L. 1965. Partial extraction of organic matter, pp. 1401-1408. IN: C. A. Black (ed.), Methods of Soil Analysis. Part 2, Chemical and Microbiological Properties, No. 9. American Society of Agronomy, Madison, WI.

6.6   Mehlich, A., 1984. Photometric determination of humic matter in soils. A proposed method. Comm. Soil Sci. Plant Anal. 15(12):1417-1422.

6.7   Strek, H. J. 1984. Improved herbicide rate recommendations using soil and herbicide property measurements. Ph.D. dissertation, Crop Sci. Dept., North Carolina State University, Raleigh, NC.

DETERMINATION OF THE RELATIVE AVAIL-ABILITIES OF HYDROGEN, CALCIUM, MAG-NESIUM, POTASSIUM, SODIUM, PHOSPHORUS, SULFUR, IRON, MANGANESE, ZINC, COPPER, ALUMINUM AND OTHER METALS BY A SMALL-EXCHANGE APPROACH TO SOIL TESTING

1. PRINCIPLE OF THE METHOD

1.1   The theoretical basis for the small-exchange approach is that the activity or effective concentrations of each ion are reflected by the concentration in solution at equilibrium when the amount exchanged (adsorbed or released) by the solid phase is negligible, compared with the total amounts adsorbed (see 12.1 to 12.4).

1.2   Ionic activities or relative partial molar free energies of ions within the soil system (ions in solution and under the influence of the adsorption force fields of the solid phase) determine their immediate availabilities or intensities within soil-water systems (see 12.1 to 12.3).

1.3   The use of DTPA to desorb and complex trace metals was proposed by Lindsay and Norvell (see 12.3, 12.5).

2. RANGE AND SENSITIVITY

2.1   Atomic absorption spectrophotometry techniques have been used to obtain the following concentration limits for ions in the filtrates without concentration or dilution. Modern inductive coupled plasma (ICP) spectrometers offer equivalent sensitivity and an opportunity to include phosphorus, boron, sulfur and molybdenum in routine testing.

Element	Range (x 10^{-4} M)	Element	Range, mg/L
Ca	4.0 to 200	Fe	0.16 to 8.0
Mg	0.8 to 40	Mn	0.16 to 8.0
K	0.2 to 10	Zn	0.06 to 3.0
Na	0.2 to 10	Cu	0.04 to 2.0
P	0.05 to 0.3	Al	0.16 to 8.0
		Cd	0.04 to 2.0
		Ni	0.04 to 2.0

2.2   Colorimetric (see 12.6) and turbidimetric (see 12.7) methods have been used with limited success to obtain an operating range for filtrates of 0.1 to 16 micromoles (μm) per liter for phosphorus and 1.0 to 40 mg/L for sulfur as sulfate, respectively, in the filtrates.

3. **INTERFERENCES**

3.1   The testing solution should be prepared fresh every week. Limited studies indicate that the relative amounts of the different elements desorbed by DTPA from soil change as the prepared solution ages.

3.2   The effect of calcium and other ions on the atomic absorption determination of cadmium is significant and requires that an adjustment be made. Background correction will eliminate this source of error.

3.3   Triethanolamine (TEA) is used to prevent excessive dissolution of trace metals from high pH soils. The buffer capacity is not sufficient, however, to prevent a reflection of soil acidity on trace element availability in acid soils.

3.4   For some soils, the colormetric determination of phosphorus is not adequate because of the development of turbidity with stannous chloride used as the reducing agent, or when phosphorus is very near the detection limit of 1 mg/L. Other reducing agents as proposed by Watanabe and Olsen (see 12.8), or the isobutyl alcohol extraction method as proposed for water soluble phosphorus (see 12.9) have been used with some success on these samples.

4. **PRECISION AND ACCURACY**

4.1   For soil samples of uniform composition, coefficients of variation ranging from 5% to 10% can be expected for determinations made by spectrophotometry.

4.2   Colormetric determinations of phosphorus and sulfur as SO_4 are not as accurate or precise as would be desired.

5. **APPARATUS**

5.1   No. 10 (2-mm) sieve, teflon or plastic.

5.2   Erlenmeyer flasks, 125-mL capacity, with hollow polyethylene stoppers.

5.3   Rotating shaker, 150 rpm for 125-mL flasks.

5.4   Filter funnels, 11 cm.

5.5   S & S No. 402S filter paper (or equivalent), 15 cm.

5.6   50 mL sample storage bottles.

5.7   pH meter with reference and glass electrodes accurate to 0.02 pH units when standardized against pH 4.0 and pH 7.0 buffers.

5.8   Precise automatic diluter for diluting soil test solutions and standards 1:50 in $SrCl_2$ solution for Mg and Ca determinations by atomic absorption spectrophotometry.

5.9   Accurate automatic diluter for diluting soil test solutions that are above the range of the working standards.

5.10   Atomic absorption spectrophotometer equipped with a background correction.

5.11   Colorimeter with 2-cm light-path cells or AutoAnalyzer for colorimetric analyses.

5.12   Analytical and top-loading balances.

5.13   50-mL test tubes, 15-mL volumetric pipets, and 1-mL measuring pipets for use in P determination.

5.14   50-mL Erlenmeyer flasks, 10-mL volumetric pipets, and 1-mL measuring pipets for S determinations.

5.15   Various beakers, Erlenmeyer flasks, graduated cylinders, volumetric pipets and flasks, and storage bottles; and a magnetic heating and stirring unit, refrigerator, and other equipment necessary for preparing and storing the stock solutions, soil test solution, reagents and standards.

6. REAGENTS

6.1   <u>Stock Solutions</u> - Prepare the following stock solutions by dissolving the indicated amount of pure salt in pure water and diluting to 1 liter.

Stock solutions	Weight of salt per liter
0.10 M NaCl	5.84 g NaCl
0.25 M KCl	18.64 g KCl
1.00 M $MgCl_2$	203.31 g $MgCl_2 \cdot 6H_2O$
2.00 M $CaCl_2$	294.04 g $CaCl_2 \cdot 2H_2O$
0.010 M KH_2PO_4	1.3609 g KH_2PO_4
1000 mg/L S	5.435 g K_2SO_4

Using the above reagents to prepare stock solutions and adding TEA results in a more constant pH than is achieved when compounds considered to provide standards of more accurate composition are used.

Prepare 1000 mg/L stock solutions of aluminum, zinc and cadmium by dissolving 1.000 g of the pure metal in 50 mL of 1:1 concentrated HCl and diluting to 1 liter with pure water. Prepare 1000 mL/L stock solutions of manganese, iron, nickel and copper by dissolving 1.000 g pure metal in 1:1 concentrated HNO_3 and diluting to 1 liter with pure water.

6.2   DTPA - (diethylenetriaminepentaacetic acid)

6.3   0.0275 M TEA - Dissolve 4.101 g triethanolamine (TEA) in pure water and dilute to 1 liter.

6.4   1% Superfloc - 127 Solution - To 700 mL pure water in a 1-liter beaker on a magnetic stirring unit, slowly add 10.0 g Superfloc-127, a small portion at a time, with continuous stirring. Cover the beaker with a watch glass and allow this solution to stir overnight. Transfer quantitatively to a 1-liter volumetric flask and dilute to 1 liter. Store this solution in the refrigerator.

6.5   Preservative Solution - Dissolve 0.1 g phenylmercuric acetate in 20 mL dioxane and dilute to 100 mL with pure water. This solution will be cloudy. Store in the refrigerator.

6.6   Small-Exchange Soil Test Solution - Dissolve 1.5734 g DTPA (see 6.2) in 300 mL pure water with gentle heating. Transfer quantitatively to a 1-liter volumetric flask. Add to this the indicated amounts of the following stock solutions.

Stock Solutions	Amount of each stock solution per liter of soil test solution
0.25 M KCl	10.0 mL
1.00 M $MgCl_2$	10.0 mL
2.00 M $CaCl_2$	25.0 mL
1% Superfloc-127 Solution	100.0 mL
Preservative Solution	10.0 mL

The solution will turn cloudy when the preservative solution (see 6.5) is added, but vigorous mixing will turn the solution clear. Dilute the flask contents to 1 liter with pure water and store the soil test solution in the refrigerator. Prepare fresh test solutions weekly.

6.7   Strontium Chloride Diluting Solution - Dissolve 161.2 g
        strontium chloride ($SrCl_2 \cdot 6H_2O$) in pure water and dilute to 1
        liter. Dilute 40.0 mL of this solution to 1 liter with pure water.

6.8   Ammonium Molybdate Reagent - Dissolve 4.17 g ammonium
        molybdate [$(NH_4)_6Mo_7O_{24} \cdot 4H_2O$] in 33 mL pure water. Add
        47 mL conc sulfuric acid (H_2SO_4). When it is cool, add 87 mL
        pure water. Store in a dark-colored glass bottle in the
        refrigerator.

6.9   Stannous Chloride Reagent - Dissolve 5 g stannous chloride
        ($SnCl_2 \cdot 2H_2O$) in 12.5 mL conc hydrochloric acid (HCl).
        Store in a dark-colored glass bottle in the refrigerator. Dilute 1
        mL of this solution to 75 mL with pure water for use on a daily
        basis.

6.10   Acid Seed Reagent - Dilute 40.0 mL 1000 mg/L sulfur standard
        to 1 liter with pure water. Mix 50 mL of this solution with 50
        mL conc hydrochloric acid (HCl) to obtain the acid seed reagent.

6.11   Barium Chloride - Obtain or prepare barium chloride ($BaCl_2 \cdot$
        $2H_2O$) crystals to pass a 20-60 mesh.

6.12   Stock Standard Solution - Prepare a combined stock standard by
        pipetting the indicated amount of each stock solution listed
        below into a 1-liter volumetric flask and diluting to volume with
        pure water.

Stock Solution	Amount of stock solution per liter of combined stock standard	Element concentration in combined stock standard
0.10 M NaCl	100 mL	10×10^{-3} M
0.25 M KCl	40 mL	10×10^{-3} M
1.00 M $MgCl_2$	40 mL	40×10^{-3} M
2.00 M $CaCl_2$	100 mL	200×10^{-3} M
1000 mg/L Al	80 mL	80 mg/L
1000 mg/L Mn	80 mL	80 mg/L
1000 mg/L Fe	80 mL	80 mg/L
1000 mg/L Ni	20 mL	20 mg/L
1000 mg/L Cu	20 mL	20 mg/L
1000 mg/L Zn	30 mL	30 mg/L
1000 mg/L Cd	20 mL	20 mg/L

6.13 Compensating Solution - Dissolve 1.573 g DTPA in 300 mL pure water. Heat briefly. Transfer this solution quantitatively to a 1-liter volumetric flask. Add to this 10.0 mL of the preservative solution and 100.0 mL of the 1% Superfloc-127 Solution and dilute to 1 liter with pure water. Store in a refrigerator.

6.14 Working Standards for Atomic Absorption Spectrophotometer - Prepare 10 flame atomic absorption standards by pipetting the indicated amount of the combined stock standard for each working standard listed below into a 500-mL volumetric flask. Add 50.0 mL Compensating Solution and 5.0 mL conc HNO_3. Dilute to volume with pure water.

Working Standard No.	Amount of combined stock standard per 500 mL working standard
1	50.0
2	40.0
3	30.0
4	20.0
5	12.5
6	8.0
7	6.0
8	4.0
9	2.0
10	1.0

Elemental Concentrations of Working Standards

Element	Unit	1	2	3	4	5
Na	10^{-4} M	10.0	8.0	6.0	4.0	2.5
K	10^{-4} M	10.0	8.0	6.0	4.0	2.5
Mg	10^{-4} M	40.0	32.0	24.0	16.0	10.0
Ca	10^{-4} M	200.0	160.0	120.0	80.0	50.0
Al	mg/L	8.0	6.4	4.8	3.2	2.0
Mn	mg/L	8.0	6.4	4.8	3.2	2.0
Fe	mg/L	8.0	6.4	4.8	3.2	2.0
Ni	mg/L	2.0	1.6	1.2	0.8	0.5
Cu	mg/L	2.0	1.6	1.2	0.8	0.5
Zn	mg/L	3.0	2.4	1.8	1.2	0.75
Cd	mg/L	2.0	1.6	1.2	0.8	0.5

Element	Unit	6	7	8	9	10
Na	10^{-4} M	1.6	1.2	0.8	0.4	0.2
K	10^{-4} M	1.6	1.2	0.8	0.4	0.2
Mg	10^{-4} M	6.4	4.8	3.2	1.6	0.8
Ca	10^{-4} M	32.0	24.0	16.0	8.0	4.0
Al	mg/L	1.28	0.96	0.64	0.32	0.16
Mn	mg/L	1.28	0.96	0.64	0.32	0.16
Fe	mg/L	1.28	0.96	0.64	0.32	0.16
Ni	mg/L	0.32	0.24	0.16	0.08	0.04
Cu	mg/L	0.32	0.24	0.16	0.08	0.04
Zn	mg/L	0.48	0.36	0.24	0.12	0.06
Cd	mg/L	0.32	0.24	0.16	0.08	0.04

6.15   Phosphorus Working Standards - Dilute 100.00 mL 0.01 M KH_2PO_4 stock solution to 1 liter with pure water to give a 0.001M P solution. Prepare phosphorus working standards in the range of 0 to 20×10^{-6} M P by diluting aliquots of the 0.001 M P standard to 100 mL with pure water. Prepare these standards for use on a daily basis. Five or six standards should be sufficient to prepare a calibration curve.

6.16   Sulfur Working Standards - To 100.0 mL 1000 mg/L sulfur standard, add 100.0 mL of the soil test solution and dilute to 1 liter with pure water to yield a 100 mg/L sulfur standard. Prepare the working sulfur standards by pipetting the indicated amounts of 100 mg/L S standard and the soil test solution for each sulfur standard listed below into a 100-mL volumetric flask and diluting to volume with pure water. Prepare these standards for use on a daily basis.

Standard no.	Amount of 100 mg/L S standard per 100 mL of work- ing S standard	Amount of soil test solution per 100 mL of work- ing S standard	S concen- tration in working S standard
1	0.00 mL	10.00 mL	0 mg/L
2	2.00 mL	9.80 mL	2 mg/L
3	5.00 mL	9.50 mL	5 mg/L
4	10.00 mL	9.00 mL	10 mg/L
5	20.00 mL	8.00 mL	20 mg/L
6	30.00 mL	7.00 mL	30 mg/L
7	40.00 mL	6.00 mL	40 mg/L

7.   PROCEDURE

7.1   Blank Preparation - Prepare a blank soil test solution by pipetting 40.0 mL pure water, 5.0 mL soil test solution, and 5.0 mL 0.0275 M TEA into a 125-mL Erlenmeyer flask and mixing. Check the pH of this solution. It should be 7.3 ± 0.05; if not, adjust the amount of 0.0275 M TEA and pure water used to obtain a pH of 7.3 ± 0.05 in a final volume of 50.0 mL of soil test solution.

7.2   Small Exchange Equilibration - Weigh 5.00 g 10-mesh soil into a 125-mL Erlenmeyer flask. Add to the flask the amount of pure water needed to obtain a pH of 7.3 ± 0.05 and a final volume of 50.0 mL with no soil present, as determined in the blank preparation step (see 7.1). Add 5.0 mL Soil Test Solution to the flask. Add to the flask the amount of 0.0275 M TEA needed to obtain a pH of 7.3 ± 0.05 and a final volume of 50 mL with no soil present as determined in the blank preparation step (see 7.1). Seal the blank and sample flasks with plastic stoppers and place the flasks on a rotating shaker for 1 hour at 150 oscillations per minute. Allow the flasks to stand for an additional 23 hours and filter the samples through S & S 402S filter paper (or equivalent). Greater precision can be achieved by working with 1 liter volumes for 20 samples at a time.

7.3   pH - Determine the pH of an aliquot of sample using a pH meter and glass and reference electrodes standardized against pH 4.0 and 7.0 buffers.

7.4   Sodium and Potassium - These elements are easily determined in the blank and soil test solution samples by flame emission spectroscopy.

7.5   Magnesium and Calcium - Dilute the standards and blank and soil test solutions samples 1:50 in the strontium chloride ($SrCl_2$) diluting solution, using a precise automatic diluter. Determine the concentration of these elements in the blank and samples by atomic absorption spectrophotometry.

7.6   Aluminum, Magnesium, Iron, Nickel, Copper, Zinc and Cadmium - Determine the concentration of these elements in the blank and soil test solution samples by atomic absorption spectrophotometry.

7.7   Phosphorus - To 15.0 mL each of blank and Soil Test Solution samples and each phosphorus standard, add 0.5 mL Ammonium Molybdate Reagent and mix. Add 0.3 mL Stannous Chloride Reagent, mix, and read the absorbencies on a colorimeter at 660 nm, using a cell with a 2-cm light path between 6 and 11 minutes.

7.8   <u>Sulfur</u> - To 10.00 mL blank and sample soil test solutions and each sulfur standard, add 1 mL Acid Seed Reagent, mix, and add 0.5 g barium chloride ($BaCl_2 \cdot 2H2$) crystals (20-60 mesh). Let stand for exactly 1 min. Mix until the crystals are dissolved and read the absorbencies on a colorimeter at 420 nm, using a cell with a 2-cm light path, between 2 and 8 minutes.

8. CALIBRATION AND STANDARDS

8.1   <u>Working Standards</u> - Working standards should be prepared as described in section 6. Samples with element concentrations above the range of the instrument or standards should be diluted with a solution containing 100 mL Compensating Solution per liter.

8.2   <u>Calibration</u> - The working standards should be used to prepare the necessary calibration curves for calculating the element concentrations in the samples. Careful attention should be paid to the instrument manufacturer's instructions in this regard. Computers greatly aid in the preparation of calibration curves and in error, analysis and quality-control programs.

9. CALCULATION

9.1   The results are reported in the units of the standards for potassium, calcium, phosphorus and magnesium. For desorbed sulfur and trace elements, the soil levels are reported (solution levels x 10). If the filtrates are diluted, the dilution factor must be applied.

9.2   Computer programs have been prepared to change values to pH, pK, pCa, pMg, pFe, pMn, etc. The validity of these conversions has been verified through the use of numerous samples from research investigations and thousands of user samples. This computerized expert system provides an intensity value and a desorbed quantity amount for most elements and relative intensity values for potassium, calcium, magnesium and hydrogen. In addition, data for Total Sorbed Metals by the U.S.E.P.A. method is included. Information regarding the computer program may be obtained from Land Management Decisions, Inc,. 1429 Harris Street, State College, PA 16803.

10. EFFECTS OF STORAGE

10.1   Soils may be stored in an air-dry condition for prolonged periods with no effect on test levels of the different elements. However, maintaining samples in a wet condition at room temperature renders the results unreliable for manganese, iron, copper, phosphorus and possibly other elements. Differences between moist and air-dried samples have been found, particularly for manganese and aluminum.

11. INTERPRETATION

11.1   Interpretative data with respect to critical levels for potassium, magnesium and phosphorus have been carried over from existing methods (see 12.1 and 12.3).

11.2   Results for samples supplied from fertility plots in several states suggest the following:

Element	Low	Normal			High
P 10^{-6}M	1	1	-	5	5
K 10^{-4}M	1.5	1.5	-	4	4
Mg 10^{-4}M	5	5	-	15	16
Ca 10^{-4}M	40	40	-	60	60
Al (ppm)	5	5	-	20	20
Mn (ppm)	20	2	-	100	100
Fe (ppm)	10	10	-	75	75
Cu (ppm)	2	2	-	40	40
Zn (ppm)	2	2	-	75	75
Cd (ppm)	0.1	0.1	-	0.5	0.5
Na (ppm)	50	50	-	150	150
S (ppm)	20	20	-	100	100

Ni, Pb, Mo, Cr not determined

Phosphorus, potassium, calcium and magnesium are expressed as concentrations in the small-exchange extract and all others are amounts desorbed from the soil (solution concentrations x 10).

11.3   The above interpretative data represent observed values for known conditions. Critical levels based on field response varies with crops and local field conditions. The low levels are not always deficient and high levels are not always toxic. The cadmium ranges listed are designed to protect the food chain (see 12.13, 12.14, 12.15, 12.16). The sensitivity of this test for molybdenum is not adequate to predict deficient levels of molybdenum in soils, but is excellent for predicting high and excessive levels associated with molybenosis in cattle (see 12.10, 12.12). Interpretations for the low end of the low range were obtained from soil samples sent from many locations within the United States (see 12.17). For other application data, see 12.18, 12.19, 12.20 and 12.21.

12. REFERENCES

12.1.   Baker, D. E., and P. F. Low. 1970. Effect of the sol-gel trans-
        formation in clay-water systems on biological activity: II.
        Sodium uptake by corn seedlings. Soil Sci. Soc. Am. Proc. 34:
        49-56.

12.2   Baker, D. E. 1971. A new approach to soil testing. Soil Sci.
        112:381-391.

12.3   Baker, D. E. 1973. A new approach to soil testing: II. Ionic
        equilibria involving H, K, Ca, Mg, Mn, Fe, Cu, Zn, Na, P and
        S. Soil Sci. Soc. Am. Proc. 27:537-541.

12.4   Baker, D. E. 1977. Ionic activities and ratios in relation to
        corrective treatments of soils, pp. 55-74. IN: Soil Testing:
        Correlating and Interpreting the Analytical Results. American
        Society of Agronomy, Madison, WI.

12.5   Viets, Jr., F. G., and W. L. Lindsay. 1973. Testing soils for
        zinc, copper, manganese and iron, pp. 153-172. IN: L. M.
        Walsh and J. D. Beaton (eds.), Soil Testing and Plant Analysis.
        Soil Science Society of America, Madison, WI.

12.6   Baker, D. E., and C. M. Woodruff. 1963. Influence of volume
        of soil per plant upon growth and uptake of phosphorus by corn
        from soils treated with different amounts of phosphorus. Soil.
        Sci. 94:409-412.

12.7   Bardsley, C. E., and J. D. Lancaster. 1960. Determination of
        reserve sulfur and soluble sulfates in soils. Soil Sci. Soc. Am.
        Proc. 24:265-268.

12.8   Watanabe, F. S., and S. R. Olsen. 1965. Test of an ascorbic
        acid method for determining phosphorus in water and $NaHCO_3$
        extracts from soil. Soil Sci. Soc. Am. Proc. 29:677-678.

12.9   Watanabe, F. S., and S. R. Olsen. 1962. Colorimetric
        determination of phosphorus in water extracts of soil. Soil
        Sci. 93:183-188.

12.10   Hornick, S. B., D. E. Baker, and S. B. Guss. 1976. Crop pro-
        duction and animal health problems associated with high soil
        molybdenum, pp. 665-684. IN: W. R. Chappel and K. K.
        Peterson (eds.), Molybdenum in the Environment.

12.11   Baker, D. E., and M. C. Amacher. 1981. Development and
        interpretation of a diagnostic soil testing program. Pa. Agric.
        Exp. Sta. Bull. 826. University Park, PA.

12.12 Murray, M. R., and D. E. Baker. 1989. Monitoring and assessment of soil and forage molybdenum near an atmospheric source. Environ. Monitoring and Assessment 12: In press.

12.13 Leach, R. M., Jr., Kathy Wei-Li Wang, and D. E. Baker. 1979. Cadmium and the food chain: The effect of dietary cadmium on tissue composition in chicks and laying hens. J. Nutr. 109 (3): 437-443.

12.14 Baker, D. E., M. C. Amacher, and R. M. Leach. 1979. Sewage sludge as a source of cadmium in soil-plant-animal systems. Environ. Health Perspectives 28:45-49.

12.15 Baker, D. E., D. S. Rasmussen, and J. Kotuby. 1984. Trace metal interactions affecting soil loading capacities for cadmium, pp. 118-132. IN: Larry P. Jackson, Alan R. Rohlik, and Richard A. Conway (eds.), Hazardous and Industrial Waste Management and Testing; Third Symposium, ASTM STP 851, American Society for Testing and Materials, Philadelphia, PA.

12.16 Baker, D. E., and M. E. Bowers. 1988. Human health effects of cadmium predicted from growth and composition of lettuce in gardens contaminated by zinc smelters, pp. 281-295. Trace Substances in Environmental Health, XXII.

12.17 Baker, D. E. 1976. Soil chemical constraints in tailoring plants to fit problem soils. I. Acid Soils. IN: Wright, M.J., (ed.), Plant Adaptation to Mineral Stress in Problem Soils. Cornell University, Ithaca, NY.

12.18 Stout, W. L., and Baker, D. E. 1978. A new approach to soil testing: III. Differential adsorption of potassium. Soil Sci. Soc. Am. J. 42:307-310.

12.19 Stout, W. L., and Baker, D. E. 1981. Effect of differential adsorption of potassium and magnesium in soils on magnesium uptake by corn. Soil Sci. Soc. Am. J. 45:996-997.

12.20 Baker, D. E., and A. M. Wolf. 1984. Soil chemistry, soil mineralogy and the disposal of solid wastes. IN: S. K. Majumdar and E. W. Miller (eds.), Solid and Liquid Wastes: Management, Methods and Socioeconomic Considerations. Pennsylvania Academy of Science, Philadelphia, PA.

12.21 Baker, D. E., and J. K. Buck. 1988. Using computerized expert systems, unique soil testing methods, and monitoring data in land management decisions. Mine Drainage and Surface Mine Reclamation, pp. 246-256.   Vol. II: Mine Reclamation, Abandoned Mine Lands and Policy Issues. U.S.D.I. Bureau of Mines Circular 9184.

DETERMINATION OF pH, SOLUBLE SALTS, NITRATE, PHOSPHORUS, POTASSIUM, CALCIUM, MAGNESIUM, SODIUM, AND CHLORIDE IN GREENHOUSE GROWTH MEDIA (SOILLESS MIXES) BY WATER SATURATION EXTRACTION

1. PRINCIPLE OF THE METHOD

1.1   Growth media (soilless mixes) used for the production of plants in greenhouses provide relatively low nutrient-holding capacity. The soil solution is the primary source of nutrient elements for plant growth. A water saturation extract of the growth media, therefore, gives a good indication of the available nutrient status. The medium is saturated with water without preliminary preparation. This procedure eliminates possible segregation of mix components and insures analysis of the growth medium as the grower is actually using it. Soluble salt and nutrient element concentrations in the water saturation extract are related to the moisture-holding capacity of the growth medium. This process eliminates the need to consider bulk density as a factor in the analysis procedure. One set of guidelines can be used with all soilless mixes (see 12.10, 12.11, 12.12, 12.13).

1.2   Water-saturation extraction for measuring the salt content of soil was adopted by the U. S. Salinity Laboratory (see 12.8). During the 1960s, Geraldson (see 12.3, 12.4) used the saturated-soil extract approach to determine the nutrient *intensity and balance* in the sandy soils of Florida. Lucas, Ricke and Doll (see 12.6) studied the saturated soil-extract method for analyzing greenhouse growth media (soilless mixes) and found it provided more meaningful results and was more advantageous than the Spurway method (see 12.7).

1.3   Water soluble levels of the key micronutrients in prepared growth media (soilless mixes) are quite low. Zinc and manganese concentrations in water saturation extracts of growth media rarely exceed 0.8 mg/l, and iron concentrations rarely exceed 4.0 mg/l. Hence, it is difficult to distinguish between deficient and adequate levels. In peat and bark-based growth media, the micronutrients are complexed by organic compounds (12.9). In evaluating 15 extractants, Berghage et al. (12.1) found that extractable levels of iron, manganese and zinc could be increased greatly by using weak solutions of various salts, acids or chelates in the saturating solution with the saturation-extract procedure. A 0.005 $\underline{M}$ DTPA was found to increase extractable micronutrient levels most consistently while having only a minor effect on the other key test parameters: total soluble salts and extractable levels of nitrate, phosphorus, potassium, calcium, magnesium, sodium and chloride.

2 . RANGE AND SENSITIVITY

2.1   This method is adapted to growth media (soilless mixes and very sandy soils) which have a relatively low nutrient-holding capability. For soils or mixes having appreciable nutrient-holding capacity, this method of analysis reflects less accurately actual nutrient element availability.

2.2   Sensitivity of analysis is dependent upon the instrumentation used. Dilution may be necessary for samples having high soluble salt levels.

3 . INTERFERENCES

3.1   pH measurements may be confounded by high soluble salt levels.

3.2   Soilless mixes containing slow-release fertilizers may give inflated results.

3.3   Interferences relevant to each analytical procedure apply.

4 . PRECISION AND ACCURACY

4.1   Reproducibility of results is dependent upon wetting the sample just to the point of complete saturation. When properly saturated, pH, soluble salt and nutrient element levels are reproducible.

5 . APPARATUS

5.1   600-mL plastic beaker.

5.2   Spatula.

5.3   Filter paper (Whatman No. 1), 11 cm.

5.4   250-mL vacuum flask,.

5.5   Buchner funnel (11 cm), fitted with appropriate rubber stopper to fit vacuum flask (see 5.4).

5.6   Vacuum pump.

5.7   Vial, snap-cap, 100 mL.

5.8   Conductivity bridge with 0 to 1 million ohm range.

5.9   Conductivity cell, dipping type, (cell constant of 1.0).

5.10 Thermometer.

5.11 pH meter with paired glass and calomel reference electrodes.

5.12 pH meter with expanded scale or specific ion meter.

5.13 Nitrate electrode with paired reference electrode.

5.14 Chloride electrode with paired reference electrode.

5.15 Colorimeter.

5.16 Flame emission and/or atomic absorption spectrophotometer.

5.17 Volumetric flasks and pipettes as required for preparation of reagents and standard solutions.

6. REAGENTS

6.1.1 Pure Water - for saturation of samples.

6.1.2 0.005 M DTPA - for saturation of samples when determination of iron, manganese and zinc is desired. For each liter of solution, weigh 1.97 g dry DTPA (Diethylenetriaminepenta - acetic acid) into a 1-liter volumetric flask. Bring to volume with 50° C pure water to facilitate dissolution. Allow it to cool to room temperature and adjust the volume.

6.2 0.01 M Potassium Chloride (KCl) - for standardizing conductivity bridge (see page 31).

6.3 Buffer Solutions - pH 4.0 and 7.0 buffers for standardizing pH meter (see page 2).

6.4 Nitrate-Nitrogen Standard (1000 mg/L) - Weigh 7.218 g potassium nitrate (KNO_3)into a 1-liter volumetric flask and bring to volume with pure water. Prepare standards containing 1, 5, 10, 50, 100 and 200 mg/L nitrate-nitrogen by diluting appropriate aliquots of the 1000 mg/L standard with pure water.

6.5 Chloride Standard (1000 mg/L) - Weigh 2.103 g potassium chloride (KCl) into a 1-liter volumetric flask and bring to volume with pure water. Prepare standards containing 1, 5, 10, 50, 100 and 200 mg/L chloride by diluting appropriate aliquots of the 1000 mg/L standard with pure water.

6.6   Phosphorus Reagents - (See appropriate section on phosphorus color development, pages 72-73).

6.7   Potassium Reagents - (See appropriate section on potassium determination, pages 82-83).

6.8   Calcium Reagents - (See appropriate section on calcium determination, pages 82-83).

6.9   Magnesium Reagents - (See appropriate section on magnesium determination, pages 82-83).

6.10 Sodium Reagents - (See appropriate section on sodium determination, pages 82-83).

7. PROCEDURE

7.1   Saturation Extraction with Pure Water

7.1.1 Fill a 600-mL beaker about 2/3 full with the growth media. Gradually add pure water (see 6.1) while mixing until the sample is just saturated. At saturation the sample will flow slightly when the container is tipped and can be easily stirred with a spatula. Depending on the growth media composition, the saturated sample may glisten as it reflects light. After mixing, allow the sample to equilibrate for one hour and then check the following criteria to ensure saturation. The saturated sample should have no appreciable free water on the surface, nor should it have stiffened. Adjust as necessary by adding either additional mix or pure water. Then allow to equilibrate for an additional half hour.

7.1.2 Determine pH on the saturated sample by carefully inserting the electrodes directly into the slury. Transfer the saturated sample to a Buchner funnel (see 5.5) lined with a filter paper (see 5.3). Be sure there is good contact between the filter paper and funnel surface by eliminating entrapped air. Insert a funnel stopper into the neck of the vacuum flask (see 5.4), apply the vacuum and collect the extract. Transfer the extract to a snap-cap vial.

7.1.3 Check the temperature of the extract and adjust the temperature dial on the conductivity bridge. Rinse the electrode, then dip the conductivity cell into the extract. Determine the electrical conductance of the extract and record in mS per cm.

7.1.4 After establishing the standard curve, determine the nitrate-nitrogen content with a nitrate electrode. Record millivolt reading on an expanded scale pH meter or specific ion meter and compare with the standard curve plotted on semilogarithmic graph paper (see 12.2).

7.1.5 After establishing the standard curve, determine the chloride content with a chloride electrode. Record the millivolt reading from an expanded scale pH meter or specific ion meter and compare with the standard curve plotted on semilogarithmic graph paper.

7.1.6 Determine the phosphorus content in an aliquot of the extract by one of the accepted colorimetric procedures (see pages 72-73).

7.1.7 Determine the potassium, calcium, magnesium and sodium content in an aliquot of the extract by flame emission or atomic absorption spectroscopy (see pages 82-83).

7.2 Saturation Extraction with 0.005 M DTPA - to improve extraction of the micronutrients.

7.2.1 Place 400 cm^3 of growth media in a 600-mL beaker. Add 100 mL of 0.005 $\underline{M}$ DTPA solution. Mix, gradually adding pure water to bring the media just to the point of saturation. See 7.1.1 for saturation criteria and equilibration time.

7.2.2 When DTPA is used in the saturation procedure, the media pH must be determined on a separate sample of the media. Use one part media by volume and two parts of pure water by volume.

7.2.3 Extract the media solution as described in 7.1.2.

7.2.4 See 7.1.3 to 7.1.7 for the determination of total soluble salts, nitrate, chloride, phosphorus and potassium, calcium, magnesium and sodium, respectively.

7.2.5 Determine the iron, manganese and zinc content in an aliquot of the DTPA extract by atomic absorption spectroscopy (see pages 148-150).

(Note: All elemental analyses except nitrate and chloride may be done through a plasma-emission spectrograph.)

8. CALIBRATION AND STANDARDS

8.1   Working Standards - Working standards should be prepared as
       indicated in section 6. If the element concentrations are outside
       the range of the instrument or standards, prepare a suitable
       dilution. Dilute only as necessary to minimize magnification of
       the error introduced by diluting.

8.2   Calibration procedures vary with instrument techniques and type
       of instrument. Take every precaution to ensure that the proper
       procedures and manufacturer recommendations are followed in the
       operation and calibration of the instruments used.

9. CALCULATION

9.1   Report soluble salt levels as mS per cm. The electrical
       conductance (mS/cm) can be converted to mg/L by multiplying
       by 640, theoretically. However, multiplying by 700 seems to
       provide a better working conversion factor (see 12.2).

9.2   Report results for nitrate-nitrogen, phosphorus, potassium,
       calcium and magnesium as mg/1 of extract.

9.3   Determine nutrient balance by calculating the percent of total
       soluble salts for each nutrient as follows:

$$\% \text{ element} = \frac{(\text{element conc}) \ (100)}{\text{total soluble salt conc}} = \frac{(\text{mg/1}) \ (100)}{(\text{mg/1})}$$

10. EFFECTS OF STORAGE

10.1   This procedure permits the extraction of moist samples just as
       they come from greenhouses. Drying of samples is unnecessary
       and undesirable. Storage of plant growth media in either dry or
       moist state will influence primarily the soluble nitrate-nitrogen
       level. When necessary, store at about 4° C.

11. INTERPRETATION

11.1   Desirable pH, soluble salt and nutrient levels vary with the
       greenhouse crop being grown and management practices. The
       following general guidelines can be used in making preliminary
       judgment of the results (12.13).

Level of Acceptance

Analysis	Low	Acceptable	Optimum	High	Very High
Soluble salt, mmho	0-0.75	0.75-2.0	2.0-3.5	3.5-5.0	5.0+
Nitrate-N mg/L	0-39	40-99	100-199	200-299	300+
Phosphorus mg/L	0-2	3-5	6-10	11-18	19+
Potassium mg/L	0-59	69-149	150-249	250-349	350+
Calcium mg/L	0-79	80-199	200+		
Magnesium mg/L	0-29	30-69	70+		

11.2   In the desired nutrient balance, the total soluble salt is comprised of the following percentages of elements: 8% nitrate-nitrogen, 12% potassium, 15% calcium and 5% magnesium. If chloride and sodium are determined, the percentage of each should be less than 10%.

11.3   The following general interpretation guidelines should be used in the DTPA extraction procedure (7.2) for boron, iron, manganese and iron. Specific desirable levels will vary with the crop being grown.

Element	Generally Adequate Range		
Boron mg/L	0.7	-	2.5
Iron mg/L	15	-	40
Manganese mg/L	5	-	30
Zinc mg/L	5	-	30

12. REFERENCES

12.1 Berghage, R. D., D. M. Krauskopf, D. D. Warncke, and I. Widders. 1987. Micronutrient testing of plant growth media: Extractant identification and evaluation. Comm. Soil Sci. Plant Anal. 18(10):1089-1110.

12.2 Dahnke, W. C. 1971. Use of the nitrate specific ion electrode in soil testing. Comm. Soil Sci. Plant Anal. 2(2):73-84.

12.3 Geraldson, C. M. 1957. Soil soluble salts -- determination of and association with plant growth. Proc. Florida State Hort. Soc. 71:121-127.

12.4 Geraldson, C. M. 1970. Intensity and balance concept as an approach to optimal vegetable production. Comm. Soil Sci. Plant Anal. 1:187-196.

12.5 Lucas, R. E., and P. E. Rieke. 1968. Peats for soil mixes. Third International Peat Congress. 3:261-263.

12.6 Lucas, R. E., P. E. Rieke, and E. C. Doll. 1972. Soil saturated extract method for determining plant nutrient levels in peats and other soil mixes. 4th International Peat Congress 3:221-230.

12.7 Spurway, C. H., and K. Lawton. 1949. Soil testing. Mich. Agr. Exp. Sta. Bul. 132.

12.8 US Salinity Laboratory Staff. 1954. Diagnosis and improvement of saline and alkali soils. Agric. Handbook No. 60, USDA, US Government Printing Office, Washington, DC.

12.9 Verloo, M. G. 1980. Peat as a natural complexing agent for trace elements. Acta Hort. 99:51 - 56.

12.10 Warncke, D. D. 1975. Greenhouse soil testing. Proc., 5th Soil-Plant Analyst Workshop, NCR-13 Comm., Bridgeton, MO.

12.11 Warncke, D. D. 1979. Testing greenhouse growing media: Update and research. Proc., 7th Soil-Plant Analyst Workshop, NCR-13 Comm., Bridgeton, MO.

12.12 Warncke, D. D. 1986. Analyzing greenhouse growth media by the saturation extraction method. HortSci 21:223-225.

12.13 Warncke, D. D., and D. M. Krauskopf. 1983. Greenhouse growth media: Testing and nutritional guidelines. Mich. State Univ. Coop. Ext. Bul. E - 1736.

EQUIVALENT FILTER PAPER TYPES

When filter paper is called for in the methods given in this hand-
book, Whatman filter papers are designed, or their equivalent.
Listed below are the equivalent papers produced by Schleicher &
Schuell.

Whatman Number	Schleicher & Schuell Number*
1	SA710
2	SA720
5	SA730
40	SA610
42	SA620

*(These equivalent numbers were provided by Brenda Lynch Marino of
  Schleicher & Schuell, and the Council wishes to thank her for providing
  this information for the handbook)